P9-DCX-935

Chemistry

Whitten

Peck

Davis

Stanley

Prepared by

Joel Caughran
University of Georgia

Charles Atwood
University of Georgia

BROOKS/COLE
CENGAGE Learning

Australia • Brazil • Japan • Korea • Mexico • Singapore • Spain • United Kingdom • United States

For product information and technology assistance, contact us at **Cengage Learning Customer & Sales Support, 1-800-354-9706**

For permission to use material from this text or product, submit all requests online at **www.cengage.com/permissions** Further permissions questions can be emailed to **permissionrequest@cengage.com**

ISBN-13: 978-0-495-39176-0
ISBN-10: 0-495-39176-X

Brooks/Cole
10 Davis Drive
Belmont, CA 94002-3098
USA

Cengage Learning is a leading provider of customized learning solutions with office locations around the globe, including Singapore, the United Kingdom, Australia, Mexico, Brazil, and Japan. Locate your local office at: **www.cengage.com/global**

Cengage Learning products are represented in Canada by Nelson Education, Ltd.

To learn more about Brooks/Cole, visit **www.cengage.com/brookscole**

Purchase any of our products at your local college store or at our preferred online store **www.ichapters.com**

Printed in the United States of America
1 2 3 4 5 6 7 12 11 10 09

CONTENTS

iii

TO THE PROFESSOR

We have used a lecture outline with tens of thousands students during the past forty years. Student response has been overwhelmingly favorable. Our colleagues who have used earlier versions of this outline have also reacted favorably. The chapter and section numbers and references to figures and tables correspond to those elements in the Whitten, Davis, Peck, and Stanley textbooks.

We have found that the use of a detailed lecture outline nearly eliminates two common problems. (1) Many students copy incessantly so they *hear* very little that the professor says, and frequently understand even less. (2) Some students make transcription errors as they take class notes. Then they wonder why the numbers and formulas they have written down don't "fit the answer" obtained by their professor. We have largely eliminated these problems by providing this detailed lecture outline that includes: (1) concise statements of important ideas, (2) important terms with brief definitions, (3) illustrative examples to be solved in class, and (4) brief statements that interpret important ideas.

We have found that when students are freed from the tedium of extensive note taking, they listen much better (and a great deal more effectively) because they know that the important ideas and terms are already written down for them. Some scribble additional notes as we talk. Our best students work ahead by filling in the Lecture Outline before class using the PowerPoint slides then write extra notes to themselves during lecture. Students can *think* about what each idea or term means as it is defined, described, and illustrated.

Perhaps the biggest advantage of using a lecture outline becomes obvious as we solve illustrative examples. Each student has a copy of the correctly stated example before him/her. Students copying problems that are to be solved in class waste no valuable class time. We have provided ample space in which solutions can be written down as the professor solves problems. Much valuable class time is gained so that more time is available for additional topics and additional depth (or breadth), as the professor chooses.

We have included material from the first twenty-eight chapters in the ninth edition of the textbook, **General Chemistry** and **General Chemistry with Qualitative Analysis**. Most of the "common core" of general chemistry courses is included in these chapters. The material is presented in the order found in the textbooks. There are references to appropriate sections, figures, and tables **in the textbook** throughout the "Lecture Outline."

You may choose to omit some of the illustrative examples. We have provided two blank numbered pages at the end of each chapter for additional notes or illustrative examples that you wish to give to your classes.

TO THE STUDENT

This outline was written to assist you in learning general chemistry. It includes concise statements of important ideas, important terms with brief definitions, illustrative examples that will be solved in class, and brief interpretative statements.

Inclusion of these materials in a lecture outline will make the study of chemistry easier and more convenient. You will be able to listen to your professor as he/she defines, explains, and describes important terms and ideas. Most are written down for you. You will not have to copy most of the classroom statements – they are included. As your professor works out illustrative examples, you may copy the solutions in the space provided.

"Lecture Outline" does not replace your textbook. Its sole purpose is to free you from the tedium associated with taking detailed notes. This facilitates more effective use of class time. There are many references to appropriate sections, figures, and tables **in your textbook** throughout this outline.

"Lecture Outline" fits into a standard three-ring binder. We suggest that you place it in one, and then break its spine so that the pages lie flat to facilitate note taking.

We welcome your suggestions for improvements in future editions of "Lecture Outline".

ACKNOWLEDGEMENTS

We are grateful to our many colleagues and former students who have used earlier editions of this lecture outline and who have made helpful suggestions. We also want to thank Joel Caughran who has assisted us with his tremendous computer skills.

One of us (CHA) would like to thank his family: Judy, Louis, and Lesley; for their perseverance over the last fourteen years as I have worked with these many projects. Judy has especially watched me struggle to finish up these projects over the last year or so. I genuinely appreciate her support, love, and guidance.

Chapter One

THE FOUNDATIONS OF CHEMISTRY

1-1 Matter and Energy

Chemistry   the science that describes matter- its properties, the changes it undergoes, and the energy changes that accompany these processes.

Matter   anything that occupies space and has mass. Everything that is tangible is matter.

Energy   the capacity for doing work. Common forms of energy include:

kinetic energy	heat energy	light energy
potential energy	electrical energy	chemical energy

Law of nature   a statement, based on experiments (observations), that is believed to be
(scientific law) true, and to which exceptions are not known.

Two of the fundamental laws of nature are the *Law of Conservation of Matter*, "there is no detectable increase or decrease in the quantity of matter during a chemical reaction or during a physical change," and the *Law of Conservation of Energy*, "energy cannot be created or destroyed in a chemical reaction or in a physical change; it can only be converted from one form to another."

The Laws of Conservation of Matter and Conservation of Energy can be combined: *The combined amount of matter and energy in the universe is fixed.* Einstein's equation gives the equivalence between matter and energy.

1

E = the amount of energy liberated

m = the mass of matter transformed into energy $E = mc^2$

c = the speed of light, 3.00×10^8 m/s

1-2   Chemistry- a Molecular View of Matter

Dalton's Atomic Theory summarized experimental observations and interpretations on the nature of atoms as known in 1808.

The basic ideas of Dalton's atomic theory can be summarized succinctly:

1. An element is composed of extremely small indivisible particles called atoms.
2. All atoms of a given element have identical properties, which differ from those of other elements.
3. Atoms cannot be created, destroyed, or transformed into atoms of another element.
4. Compounds are formed when atoms of different elements combine with each other in small whole-number ratios.
5. The relative numbers and kinds of atoms are constant in a given compound.

An **atom** is the smallest particle of an element that maintains its chemical identity through all chemical and physical changes.

Fundamental particles are the basic building blocks of atoms, they consist of electrons, protons, and neutrons.

A **molecule** is the smallest particle of an element or compound that can have a stable independent existence.

1-3   States of Matter

We arbitrarily classify matter into three states because this enables us to deal with large amounts of information in a systematic way.

The solid state   Solids are rigid; they have definite shapes, and the volumes of solids are very nearly independent of temperature and pressure.

The liquid state   Liquids flow and assume the shapes of their containers. Liquids are only very slightly compressible and, for all practical purposes, the volume of a liquid is fixed.

2

The gaseous state (the vapor state)	Gases assume the shapes of their containers and fill completely any container in which they are placed. Gases are easily compressed and they expand indefinitely.

The above statements indicate the following. In the *solid state* the individual particles (ions, atoms, or molecules) occupy fixed positions. They are in close contact with each other. The attractive forces among the individual particles are strong. In the *liquid state* the individual particles are in contact with each other. The fact that liquids flow indicates that the attractive forces among the individual particles are not so strong as in the solid state. In the *gaseous state* the molecules are quite far apart, and the attractive forces among molecules are very minimal. See Figure 1-7.

1-4   Chemical and Physical Properties

Properties	characteristics of matter. We conveniently divide the properties of matter into two classes, chemical and physical properties.
Chemical properties	properties that matter can exhibit *only* by undergoing a change in composition, i.e., a chemical change.
Physical properties	properties that matter exhibits without undergoing a change in composition. Color, hardness, physical state, melting point, and boiling point are physical properties. See Table 1-2.
Extensive properties	properties that depend on the amount of material examined.
Intensive properties	properties that do not depend on the amount of material examined.

1-5   Chemical and Physical Changes

Chemical changes	changes in which the composition of matter changes. Iron rusts when it is exposed to moist air. This is a chemical change - a chemical property of iron is exhibited as it rusts.

3

Physical changes	changes in which the composition of matter does not change. Common physical changes include changes of state, e.g., ice melts, i.e., solid water changes to liquid water. Water boils, i.e., liquid water changes to steam (gaseous water). See Figure 1-10.

1-6   Mixtures, Substances, Compounds, and Elements

Mixtures	combinations of two or more pure substances. Each substance retains its composition and properties in a mixture. A mixture may be *homogeneous* (a *solution*) or *heterogeneous*.
Substances	any kind of matter, all samples of which have identical composition *and* identical physical and chemical properties. A substance cannot be further broken down or purified by physical means.
Compounds	substances that are composed of two or more elements in a definite ratio of mass. Compounds can be decomposed into elements by chemical changes. See Figure 1-14.
Elements	substances that cannot be decomposed into simpler substances by chemical reactions. Elements are the simplest forms of matter.

The *Law of Definite Proportions* states that the different samples of any pure compound contain the same elements in the same proportions by mass.

Abundances of a few common elements are listed. Each element is represented by a *symbol*. Symbols for elements should be learned as they are encountered. See Table 1-3.

Abundance of Some Elements in the Earth's Crust, Oceans, and Atmosphere

Oxygen, O	49.5%	Calcium, Ca	3.4%
Silicon, Si	25.7%	Sodium, Na	2.6%
Aluminum, Al	7.5%	Potassium, K	2.4%
Iron, Fe	4.7%	Magnesium, Mg	1.9%

all other elements 2.3%

4

1-7   Measurements in Chemistry

In 1964 the International System of Units (SI units) was adopted by the National Bureau of Standards. (Now the National Institute of Standards and Technology.) Of necessity, a change of units occurs over a long period of time. The metric system was "legalized" in this country in 1866! The SI system is based on the following *seven fundamental units*. Other units are derived from them. See Table 1-4.

FUNDAMENTAL SI UNITS

Physical Property	Name of Unit	Symbol
Length	Meter (39.37 in)	m
Mass	Kilogram (2.205 lb)	kg
Time	Second	s
Electric current	Ampere	A
Temperature (Thermodynamic)	Kelvin	K
Luminous intensity	Candela	cd
Amount of a substance	Mole	mol

In science, measurements are usually expressed in the units of the SI (metric) system. It is a decimal system, i.e., each unit is related to other units by a power of ten. A series of prefixes is used to denote *powers of ten of units of measurement*. See Table 1-6.

SOME COMMON PREFIXES

Prefix	Abbreviation	Means
mega	(M)	1,000,000 times (10^6)
kilo	(k)	1000 times (10^3)
deci	(d)	0.10 times <u>or</u> 0.10 of (10^{-1})
centi	(c)	0.010 times <u>or</u> 0.010 of (10^{-2})
milli	(m)	0.0010 times <u>or</u> 0.0010 of (10^{-3})
micro	(μ)	0.0000010 times <u>or</u> 0.0000010 of (10^{-6})
nano	(n)	0.0000000010 times <u>or</u> 0.0000000010 of (10^{-9})
pico	(p)	0.0000000000010 times <u>or</u> 0.0000000000010 of (10^{-12})

1-8   Units of Measurement

Mass   A measure of the quantity of matter a body contains. Mass is an invariable quantity. The basic unit of mass in the SI system is the *kilogram*, the mass of a platinum-iridium alloy stored near Paris, France. 1 kilogram (kg) is 2.205 pounds. See Tables 1-7 and 1-8.

Weight   A measure of the earth's gravitational attraction for a body. Weight varies with distance from the center of the earth.

Length   The *meter* is the standard unit of length (distance) in the SI system. One meter is the distance light travels in a vacuum in 1/299,792,468 second. (In 1 second, light travels 299,792,468 meters in a vacuum.) 1 meter is approximately 39.37 inches. See Figure 1-17 and Table 1-8.

Volume   The common units of volume are the liter and the milliliter (cubic centimeter) in the metric system. The SI unit is the cubic meter, a very large volume. See Figure 1-18 and Table 1-8.

It is helpful to remember one equivalence between English units and SI and metric units. Refer to Appendix C and Table 1-8.

Mass and Weight   1 pound = 453.6 grams (usually rounded to 454 g) at sea level

Length   1 meter = 39.37 inches   *or*   1 inch = 2.54 cm (exactly)

Volume   1 liter   = 1.057 quarts   *or*   1 quart = 0.946 liter

1-9   Use of Numbers

Chemistry is a quantitative science in which we measure and calculate many things. We must, therefore, be able to use numbers. We need to know how to represent very large and very small numbers *and* we must indicate how accurately we actually know the numbers being used.

Scientific Notation

When we deal with very large and very small numbers we use **scientific notation.** Consider 108 grams of silver contains approximately

6

602,000,000,000,000,000,000,000 silver atoms   represented in scientific notation as

6.02×10^{23} silver atoms.

The mass of one silver atom is approximately

0.000 000 000 000 000 000 000 179 gram   represented in scientific notation as

1.79×10^{-22} gram

With such numbers, it is inconvenient to write down all of the zeroes. In scientific notation, we use powers of ten (exponents) to replace the zeroes. We must place one *nonzero* digit to the left of the decimal.

$7,400,000 = 7.4 \times 10^6$ 6 places to the left, $\therefore$ the exponent of 10 is 6

$0.000246 = 2.46 \times 10^{-4}$ 4 places to the right, $\therefore$ the exponent of 10 is -4

The reverse process converts numbers from exponential to decimal form.

Significant Figures

Review Appendix A for Significant Figures and Scientific Notation

Exact numbers are exact numbers, i.e., no estimation is involved. They are obtained by counting or from definitions. Most numbers are not exact - most numbers represent measurements.

Significant figures are digits believed to be correct by the person making the measurement. This assumes competence on the part of the person making the measurement. Consider the following example.

Suppose one measures a volume of water in a 10 mL graduated cylinder and reports the volume as 4.73 mL. What does this number mean? A 10 mL graduated cylinder is calibrated in tenths of milliliters.

Suppose one reads in a newspaper that the population of a city is 2,678,342 people. What does this number mean?

Accuracy   describes how closely measured values agree with correct values
Precision   describes how closely individual measurements agree with each other

The last significant figure is always the first estimated digit.

1.   Nonzero digits are always significant.

2.   Zeroes are sometimes significant, and sometimes they are not.

   a. Zeros at the *beginning* of a number (used to place the decimal) are never significant

   b. Zeros *between* nonzero digits are always significant.

   c. Zeros at the end of a number that contains a decimal point are always significant.

   d. Zeroes at the end of a number that does not contain a decimal point may or may not be significant.

3.   Exact numbers (including defined quantities) have unlimited number of significant figures.

4.   In the case of addition and subtraction, the last digit retained in the sum or difference is determined by the position of the first doubtful digit.

   In these examples the last significant digit is underlined.

$$
\begin{array}{r}
3.692\underline{3} \\
+1.23\underline{4} \\
+2.0\underline{2} \\
\hline
\end{array}
\qquad\qquad
\begin{array}{r}
8.739\underline{7} \\
-2.12\underline{3} \\
\hline
\end{array}
$$

5.   In multiplication and division, an answer contains no more significant figures than the least number of significant figures used in the operation.

   In these examples the last significant digit is underlined.

$$
\begin{array}{r}
4.24\underline{2} \\
\times\ \ 1.2\underline{3} \\
\hline
\end{array}
\qquad\qquad
\begin{array}{r}
2.783\underline{2} \\
\times\ \ \ \ 1.\underline{4} \\
\hline
\end{array}
$$

8

Rounding Off

When the number to be dropped is less than 5, the preceding digit is left unchanged.

When the number to be dropped is greater than 5, the preceding digit is increased by 1.

When the number to be dropped is 5, the preceding number is left unchanged if it is *even*; it is increased by 1 if the preceding number is odd.

1-10   The Unit Factor Method (Dimensional Analysis)

Problem-solving usually involves changing one set of units to another set of units. It transforms information from one form into another (frequently more useful) form. We shall use the "unit factor method", sometimes called "dimensional analysis". Before we begin to solve problems let us think for a moment about what various units mean. What is one foot? One foot is the distance that the National Institute for Standards and Technology defines it to be. In the U.S.A., **every unit** *is whatever the National Institute for Standards and Technology defines it to be*.

Consider the following relationship. One foot is equal to 12 inches. This relationship can be converted into two *unit factors*. They are unit factors because the numerator and denominator describe the same distance.

Similarly, one pound is 16 ounces. We can convert this information into two *unit factors*. They are unit factors because the numerator and denominator describe the same weight or mass.

One must not confuse the number combination generated by an electronic calculator with "the answer", as the following examples demonstrate. These examples illustrate "how" problems are solved, i.e., the unit factor technique (dimensional analysis). See Table 1-8 and Appendix C for conversion factors.

Example 1-1:   Express 9.43 yards (yd) in millimeters (mm).

9

Example 1-2:   Express 627 milliliters (mL) in gallons (gal).

Example 1-3:   Express 2.61×10^4 square centimeters (cm^2) in square feet (ft^2).

Example 1-4:   Express 2.61 cubic feet (ft^3) in cubic centimeters (cm^3).

A unit factor raised to any real power is still a unit factor.

1-11 Percentage

Percentage is parts per hundred of a sample. If one hundred grams of an aluminum alloy contains 87 grams of aluminum, we say that the alloy is 87 percent Al.

$$\frac{87 \text{ g Al}}{100 \text{ g sample}} \times 100\% = 87\% \text{ Al}$$

Example 1-5:   A 335 g sample of ore yields 29.5 g of iron. What is the percent of iron in the ore?

1-12 Density and Specific Gravity

Density   mass per unit volume. Symbolically, $D = \dfrac{M}{V}$. See Table 1-9.

Example 1-6:   Calculate the density of a substance if 742 grams of it occupies 97.3 cubic centimeters.

10

Example 1-7:   Suppose you need 125 grams of a corrosive, volatile liquid for a reaction. What volume of liquid would you use? The density of the liquid is 1.32 g/mL.

Specific gravity   the ratio of the density of a substance to the density of water, both at the same temperature.

$$\text{Specific Gravity} = \frac{\text{Density (substance)}}{\text{Density (water)}}$$

The density of water is very nearly 1.00 g/mL at room temperature (also 1.00 g/cm^3). The specific gravity (which has no units) of a substance is equal to its density in g/mL near room temperature.

Example 1-8:   A 31.0-gram piece of chromium is dropped into a graduated cylinder that contains 5.00 mL of water. The water level rises to 9.32 mL. What is the specific gravity of chromium?

Example 1-9:   A concentrated hydrochloric acid solution is 36.31% HCl and 63.69% H$_2$O by mass. Its specific gravity is 1.185. What mass of pure HCl is contained in 175 mL of this solution?

1-13 Heat and Temperature

Temperature refers to the intensity of heat in a body, that is, how hot or how cold the body is. Three common temperature scales are used. Therefore it is necessary to be familiar with all three and to know the relationships that exist among them. By international agreement the freezing point and boiling point of pure water at 1 atmosphere of pressure (760 torr) are reference points for some temperature scales. See Figure 1-22.

_____ bp of H$_2$O (1 atm)

_____ mp of ice (1 atm)

11

Note that on the three temperature scales the freezing point of water is 32°F, 0° Celsius, and 273 kelvins. The boiling point of water is 212°F, 100°C and 373 kelvins. In each case the three temperature measurements represent the same temperature. (The ° sign is not used with the Kelvin scale.)

Example 1-10: Convert 211°F to degree Celsius.

Example 1-11: Express 548 K in Celsius degrees.

1-14   Heat Transfer and the Measurement of Heat

A chemical reactions and physical changes occur with either an *exothermic process* (evolution of heat) or *endothermic process* (absorption of heat).

We shall use the primary SI unit of heat and energy, the joule (J), which is defined as 1 kg m^2/s^2. Another unit, the calorie, is defined as exactly 4.184 J. Historically, the calorie was defined as the amount of heat required to raise the temperature of one gram of water from 14.5°C to 15.5°C.

Specific heat

The specific heat of a substance is the amount of heat (usually expressed in joules) required to raise the temperature of one gram of the substance one degree Celsius with no change in state. Frequently, the specific heat of a substance is expressed on a per mole basis. This is the amount of heat required to raise the temperature of *one mole* of a substance one degree Celsius with no change in state, and is referred to as its *heat capacity*. Specific heats for a few common substances are tabulated below. (Appendix E has more.) Note that the specific heat of water is very high.

Specific Heats of a Few Common Substances

Substance	J/g°C	cal/g°C
ice	2.09	0.500
liquid water	4.18	1.00
steam	2.03	0.484
iron	0.444	0.106
mercury	0.138	0.0331

Example 1-12:   Calculate the amount of heat required to raise the temperature of 200.0 grams of water from 10.0°C to 55.0°C.

Example 1-13:   Calculate the amount of heat required to raise the temperature of 200.0 grams of mercury from 10.0°C to 55.0°C.

NOTE: In Examples 1-12 and 1-13, $\Delta T = 45°C$ for both H_2O and Hg

The specific heat of H_2O is large (4.18 J/g°C)

The specific heat of Hg is small (0.138 J/g°C)

Therefore, much more heat is required to raise the temperature of 200.0 grams of water by 45.0°C than for 200.0 grams of mercury

Synthesis Question

It has been estimated that 1.0 g of seawater contains 4.0 pg of Au. The total mass of seawater in the oceans is 1.6×10^{12} Tg, 1.0 Tg = 10^{12} g. If all of the gold in the oceans were extracted and spread evenly across the state of Georgia, which has a land area of 58,910 mile^2, how tall, in feet, would the pile of Au be? Density of Au is 19.3 g/cm^3.

Group Activity

On a typical day, a hurricane expends the energy equivalent to the explosion of two thermonuclear weapons. A thermonuclear weapon has the explosive power of 1.0 million tons of nitroglycerin. Nitroglycerin generates 7.3 kJ of explosive power per gram of nitroglycerin. A hurricane is energy that comes from the evaporation of water that requires 2.3 kJ per gram of water evaporated. How many gallons of water does a hurricane evaporate per day?

Chapter Two

CHEMICAL FORMULAS AND COMPOSITION STOICHIOMETRY

We shall discuss the symbolism and language used to describe matter and the changes in composition that matter undergoes. Formulas and equations are used to describe chemical reactions. The table inside the front cover of your text contains a list of the known elements, their symbols, and their atomic weights. Refer to it as necessary.

Stoichiometry describes the quantitative relationships among elements in compounds and among substances as they undergo chemical changes.

2-1   Chemical Formulas

Formulas for Elements

The **chemical formula** for a substance shows its chemical composition. By composition we mean the elements present and the ratio in which atoms of the elements occur in the substance.

Monatomic molecules   _____
(elements)

Diatomic molecules   _____
(elements)

More complex molecules   _____
(elements)

17

Formulas for Compounds

A compound contains two or more elements in chemical combination in a definite ratio by mass. The formula for a compound indicates (1) the elements present *and* (2) the ratio in which atoms of the elements occur. See Table 2-1.

Compound	1 Molecule Contains
HCl	_____
H_2O	_____
NH_3	_____
C_3H_8	_____

The *Law of Definite Proportions (Constant Composition)* is based on the observation that "Different pure samples of a compound always contain the same elements in the same proportion by mass; this corresponds to atoms of these elements combined in fixed numerical ratios." A few "exceptions" are known and these will be described later.

The structural formula not the chemical formula shows the order in which the atoms are connected.

2-2   Ions and Ionic Compounds

Ion   an atom or group of atoms that carries an electrical charge.

Anions   negatively charged ions. See Table 2-2.

Cations   positively charged ions. See Table 2-2.

Ionic compounds   consist of an extended array of ions in which the total positive and negative charges are equal. See Figure 2-2.

Formula Unit   of an ionic compound, the simplest whole number ratio of ions in the compound.

Polyatomic ions   are groups of atoms that have an electric charge

18

2-3   Names and Formulas of Some Ionic Compounds

During your study of chemistry you will often have occasion to refer to compounds by name. This section provides a brief introduction to naming a few compounds.

Tables 2-2 contains the names of some common ions. You should learn the names and formulas of these frequently encountered ions. These ions can be used to write the formulas and names of many ionic compounds. The formula of an ionic compound is written by adjusting the relative numbers of positive and negative ions so that the total charges sum to zero. The compound's name is given by giving the names of the ions, with the positive ion given first.

NaCl is named sodium chloride   (Na^+ Cl^-).

The name of H_2SO_4 is _____

The name of C_2H_5OH is _____

The formula of nitric acid is _____

The formula of sulfur trioxide is _____

The name of $FeBr_3$ is _____

The name of K_2SO_3 is _____

The charge on the sulfite ion is _____

The formula of ammonium sulfide is _____

The charge on the ammonium ion is _____

The charge on the sulfide ion is _____

The formula of aluminum sulfate is _____

The charge on the aluminum ion is _____

The charge on the sulfate ion is _____

2-4   Atomic Weights

The *atomic weight* (atomic mass) of an element is the weighted average of the masses of its constituent isotopes. A working definition of atomic weights is: atomic weights are the relative masses of atoms of different elements that are proportional to the actual masses of atoms.

19

Element	Atomic Weight	Element	Atomic Weight
H	_____	Mg	_____
N	_____	Ag	_____

2-6   The Mole

Atoms and molecules are so very small that we commonly use a unit called *the mole*, which is 6.022×10^{23} formula units of the substance under consideration. Atoms, ions, and molecules have specific masses. In addition to being 6.022×10^{23} formula units of a substance, one mole is also a specific mass of a substance. The number 6.022×10^{23} is called *Avogadro's Number*. The abbreviation for mole is mol.

The mass of one mole of atoms of a pure element in grams is numerically equal to the atomic weight to that element in atomic mass units which is also called the molar mass.

The *symbol* for an element can (1) identify the element, (2) represent one atom of the element, or (3) represent one mole of a monatomic element such as helium or neon. One mole of atoms of an element is 6.022×10^{23} atoms of the element, and has a mass in grams numerically equal to the atomic weight of the element.

One Mole of Atoms of Some Common **Elements**

Element	1 mole of atoms	Contains
H	_____ g hydrogen	_____ H atoms
Na	_____ g sodium	_____ Na atoms
Ti	_____ g titanium	_____ Ti atoms

The atomic weight of an element and Avogadro's Number can be used to relate the masses of samples of elements to the number of atoms in the sample. Consider the metal magnesium.

Example 2-1:   Calculate the mass of a iron atom in grams to three significant figures. (Because atoms are so small their masses should also be small, typically $\sim 10^{-23}$ g.)

20

Example 2-2:   Calculate the number of atoms in one-millionth of a gram of magnesium to three significant figures.

Example 2-3:   How many atoms are contained in 1.67 moles of magnesium?

Example 2-4:   How many moles of magnesium atoms are present in 73.4 g of magnesium?

2-7   Formula Weights, Molecular Weights, and Moles

The formula weight (mass) of a substance is the sum of the atomic weights of the elements in the formula, each taken the number of times the element occurs. Formula weights (masses), like the atomic weights on which they are based, are relative masses.

The formula weight of:

propane, C_3H_8, is   calcium nitrate, $Ca(NO_3)_2$, is

21

One mole of a substance is 6.022×10^{23} formula units of the substance as well as a definite mass of the substance. It is the number of grams numerically equal to the formula weight of the substance.

One mole of propane is _____ grams of C_3H_8.

One mole of calcium nitrate is _____ grams of $Ca(NO_3)_2$.

One mole of any *molecular* substance is 6.022×10^{23} *molecules* of the substance.

One Mole of Some Common **Molecular** Substances

Substance	1 Mole	Contains	
Cl_2 chlorine	____ g	_____	Cl_2 molecules
		_____	Cl atoms
C_3H_8 propane	____ g	_____	C_3H_8 molecules
		_____	C atoms
		_____	H atoms

Ionic compounds exist as combinations of discrete ions at reasonable temperatures and pressures. There are no molecules of ionic compounds near room temperature. One mole of an ionic compound contains 6.022×10^{23} *formula units* of the compound.

One Mole of Some Common **Ionic** Compounds

Compound	1 Mole	Contains	
LiCl	____ g	_____	Li^+ ions
		_____	Cl^- ions
$Ca(NO_3)_2$	____ g	_____	Ca^{2+} ions
		_____	NO_3^- ions

The following examples illustrate the ideas we have just developed.

Example 2-5:   Calculate the number of C_3H_8 molecules in 74.6 g of propane.

Example 2-6:   What is the mass of 10.0 billion propane molecules?

Example 2-7:   How many (a) moles, (b) molecules, and (c) oxygen atoms are contained in 60.0 grams of ozone, O_3? The layer of ozone in the stratosphere is very beneficial to life on earth.

a.

b.

c.

Example 2-8:   Calculate the number of oxygen atoms in 26.5 grams of lithium carbonate, Li_2CO_3, a compound used to treat manic psychosis.

A *millimole* (mmol) is 1/1000 of a mole. The relationship between moles and millimoles is illustrated in the following table.

Comparison of Moles and Millimoles

Compound	1 mol	1 mmol		
C_3H_8	_____ g	_____ g	or	_____ mg
$(COOH)_2$	_____ g	_____ g	or	_____ mg

Example 2-9: Calculate the number of millimoles in 0.234 gram of oxalic acid, $(COOH)_2$.

2-7 Percent Composition and Formulas of Compounds

If the formula of a compound is known, its chemical composition can be expressed as the percent by mass of each element in the compound. For example, one propane molecule, C_3H_8, contains *three* carbon atoms and *eight* hydrogen atoms. One mole of C_3H_8 contains *three* moles of carbon atoms and *eight* moles of hydrogen atoms. Symbolically, the percent composition of propane can be represented as:

All samples of pure propane have this composition.

Example 2-10: Calculate the percent composition of iron(III) sulfate, $Fe_2(SO_4)_3$, which is used as a mordant in textile dying.

All sample of pure iron(III) sulfate have this composition.

24

2-8 Derivation of Formulas from Elemental Composition

Thousands of new compounds are made and discovered each year. The *simplest* (or *empirical*) *formula* for a compound is the *smallest whole number ratio of atoms* present in the compound. The *molecular* (or *true*) formula for a compound is the *actual ratio of atoms* present in a compound; it may be the same as or a multiple of the simplest formula. Simplest formulas can be calculated from mass composition or percent composition data as the following examples illustrate.

Example 2-11:   A compound contains 24.74% potassium, 34.76% manganese, and 40.50% oxygen by mass. What is its simplest formula?

Example 2-12:   A sample of a compound contains 6.541 g of cobalt and 2.368 g of oxygen. What is the simplest formula for the compound?

These problems demonstrate that the ratio of atoms in the simplest formula for a compound is the same as the ratio of moles of atoms of the elements in a sample of the compound.

2-9   Determination of Molecular Formulas

   If the simplest formula for a compound *and* its formula weight are known, the molecular formula for the compound can be determined as Example 2-13 demonstrates.

Example 2-13:   A compound is found to contain 85.63% carbon and 14.37% hydrogen by mass. In another experiment its molecular weight is found to be 56.1 g/mol. What is its molecular formula?

The Law of Multiple Proportions

   When two elements, A and B, form more than one compound, the ratio of the masses of element B that combine with a given mass of element A in each of the compounds can be expressed by small whole numbers.

Example 2-14:   Show that the compounds NO_2 and N_2O_5 obey the law of multiple proportions.

2-10 Some Other Interpretations of Chemical Formulas

We can use the mole concept and the meanings of chemical formulas in many ways.

Example 2-15:   What mass of phosphorous is contained 45.3 grams of $(NH_4)_3PO_4$?

Example 2-16:   What mass of ammonium phosphate, $(NH_4)_3PO_4$, would contain 15.0 grams of nitrogen?

Example 2-17:   What mass of propane, C_3H_8, contains the same mass of carbon as is contained in 1.35 grams of barium carbonate, $BaCO_3$?

2-11 Purity of Samples

The *percent purity* is the mass percentage of a specified substance in an impure sample.

Example 2-18: A bottle of sodium phosphate, Na_3PO_4, is 98.3% pure Na_3PO_4. What are the masses of Na_3PO_4 and impurities in 250 grams of this sample of Na_3PO_4?

Synthesis Question

In 1986, Bednorz and Muller succeeded in making the first of a series of chemical compounds that were superconducting at relatively high temperatures. This first compound was La_2CuO_4 which superconducts at 35K. In their initial experiments, Bednorz and Muller made only a few mg of this material. How many La atoms are present in 3.56 mg of La_2CuO_4?

28

Group Question

Within a year after Bednorz and Muller's initial discovery of high temperature superconductors, Wu and Chu had discovered a new compound, $YBa_2Cu_3O_7$, that began to superconduct at 100 K. If we wish to make 1.00 pound of $YBa_2Cu_3O_7$, how many grams of yttrium must we buy?

Chapter Three

CHEMICAL EQUATIONS AND REACTION STOICHIOMETRY

Chemical equations represent a very precise, yet a very versatile language.

3-1   Chemical Equations

Chemical equations are used to represent (describe) chemical reactions. An equation shows (1) the substances that react, *the reactants*, (2) the substance that are formed in the reaction, *the products*, and (3) *the relative amounts* of all substances involved in the reaction. See Figure 3-1.

As a typical example, consider the reaction of ionic iron(III) oxide, Fe_2O_3, with carbon monoxide. Both CO and CO_2 are molecular compounds. The equation tells us

$$Fe_2O_3 \ + \ 3\,CO \ \xrightarrow{\Delta} \ 2\,Fe \ + \ 3\,CO_2$$

qualitatively

atomic and molecular level

molar level

masses

32

The *Law of Conservation of Matter* states, "There is no detectable change in the quantity of matter during an ordinary chemical reaction." This law provides the basis for balancing chemical equations as well as for calculations based on chemical equations.

A balanced chemical equation must always include the same number of each kind of atom on both sides of the equation.

1. Propane, C_3H_8, burns in oxygen to give carbon dioxide and water.

2. Ammonia, NH_3, burns in oxygen to give nitrogen oxide, NO, and water.

3. Heptane burns in oxygen to give carbon dioxide and water.

The *Law of Definite Proportions* states, "Different samples of a pure compound always contain the same elements in the same proportions by mass." Formulas represent *experimentally determined ratios of elements*. They must be written correctly and they must not be changed as one attempts to balance equations.

3-2 Calculations Based on Chemical Equations

We are now ready to use chemical equations to calculate the relative amounts of substances involved in chemical reactions. Consider a reaction that occurs in blast furnaces in the production of iron. Iron(III) oxide reacts with carbon monoxide to form iron and carbon dioxide. We may interpret this equation on a molecular basis.

$$\underset{\text{iron(III) oxide}}{Fe_2O_3} \quad + \quad \underset{\text{carbon monoxide}}{3\,CO} \quad \xrightarrow{\Delta} \quad \underset{\text{iron}}{2\,Fe} \quad + \quad \underset{\text{carbon dioxide}}{3\,CO_2}$$

Example 3-1: How many carbon monoxide molecules are required to react with 25 formula units of Fe_2O_3?

33

Example 3-2:   How many iron atoms can be produced by the reaction of 2.50×10^5 formula units of iron(III) oxide with excess carbon monoxide?

Formulas can also represent moles of substances involved in chemical reactions. The equation we just considered can also be interpreted on a *"per mole"* basis.

$$Fe_2O_3 \quad + \quad 3\,CO \quad \xrightarrow{\Delta} \quad 2\,Fe \quad + \quad 3\,CO_2$$

Chemical equations define *reaction ratios*, i.e., the mole ratios of reactants and products, as well as the relative masses of reactants and products.

Example 3-3:   What mass of carbon monoxide is required to react with 146 grams of iron(III) oxide? Note that this is a calculation involving both reactants.

From a balanced equation, we can construct many unit factors. For example, we can construct two unit factors that relate the *number of moles* of *any two reactants and/or products* for this particular reaction.

We can construct unit factors that relate the *masses* of any two reactants and/or products.

Example 3-4:   What mass of carbon dioxide can be produced by the reaction of 0.540 mole of iron(III) oxide with excess carbon monoxide? Note that this is a calculation involving one reactant and one product.

Example 3-5:   What mass of iron(III) oxide reacted with excess carbon monoxide if the carbon dioxide produced by the reaction had a mass of 8.65 grams? Note that this is a calculation involving one reactant and one product.

Example 3-6: How many pounds of carbon monoxide would react with 125 pounds of iron(III) oxide? Note that this is a calculation involving one reactant and one product.

We have used some of the unit factors that relate amounts of Fe_2O_3 and CO that are chemically equivalent. Let's write down nine of them. The reciprocal of each is also a unit factor, so we see that there are 18 units factors that relate Fe_2O_3 and CO. **This true for any two reactants and/or products in any chemical equation.**

3-3   The Limiting Reactant Concept

A kitchen example of the limiting reactant concept.

1 packet of muffin mix + 2 eggs + 1 cup of milk $\rightarrow$ 12 muffins

How many muffins can we make with the following amounts of mix, eggs, and milk?

Mix packets	Eggs	Milk	# of Muffins	Limiting Reactant
1	1 dozen	1 gallon	_____	_____
2	1 dozen	1 gallon	_____	_____
3	1 dozen	1 gallon	_____	_____
4	1 dozen	1 gallon	_____	_____
5	1 dozen	1 gallon	_____	_____
6	1 dozen	1 gallon	_____	_____
7	1 dozen	1 gallon	_____	_____

In the problems worked so far we have assumed the presence of an excess of one reactant. These calculations were based on the substance present in *lesser amount*, i.e., the calculations were based on the *limiting reactant*. Consider the following examples.

Example 3-7: Suppose a box contains 87 bolts, 110 washers and 99 nuts. How many sets, each consisting of one bolt, two washers, and one nut, can you construct from the contents of the box?

Example 3-8:   What is the maximum mass of sulfur dioxide that can be produced by the reaction of 95.6 g of carbon disulfide with 110 g of oxygen?

$$CS_2 \quad + \quad 3\,O_2 \quad \xrightarrow{\Delta} \quad CO_2 \quad + \quad 2\,SO_2$$

3-4   Percent Yields from Chemical Reactions

Many chemical reactions do not go to completion. That is, all reactants are not converted completely to products. The term *percent yield* is used to indicate how much of a desired product is obtained from a particular "run" of a reaction. Symbolically, percent yield may be represented as

So far the masses we have calculated from chemical equations were based on the assumption that each reaction occurred to the extent of 100%. The *theoretical yield* is the yield calculated assuming 100% reaction *and* isolation of 100% of the desired product. The *actual yield* is the yield **actually obtained** from a particular "run" of a reaction.

Example 3-9:   A 10.0 g sample of ethanol, C_2H_5OH, was boiled with excess acetic acid, CH_3COOH, to produce 14.8 g of ethyl acetate, $CH_3COOC_2H_5$ ($C_4H_8O_2$). What is the percent yield of ethyl acetate?

$$CH_3COOH \quad + \quad C_2H_5OH \quad \xrightarrow{\Delta} \quad CH_3COOC_2H_5 \quad + \quad H_2O$$

36

3-5   Sequential Reactions

There are many reactions in which the reactants have been prepared by previous reactions. Reactions of two or more steps are called *sequential reactions*. An example is the preparation of aniline from benzene.

Example 3-10:   Starting with 10.0 g of benzene (C_6H_6), calculate the theoretical yield of nitrobenzene ($C_6H_5NO_2$) and of aniline ($C_6H_5NH_2$).

If 6.7 g of aniline is prepared from 10.0 g of benzene, what is the percentage yield?

3-6   Concentrations of Solutions

We frequently carry out reactions in which one or more of the reactants is in solution. A *solution* is a homogeneous mixture consisting of a *solvent* - the dispersing (dissolving) medium, usually a liquid, and one or more *solutes* - the dissolved species. The "strength" of a solution is expressed in terms of its concentration.

Concentration   the amount of solute dissolved in a given amount of solution *of a given* amount of solvent.

Dilute solutions   solutions that contain relatively small amounts of solute in relatively large amounts of solvent (or solution).

37

Concentrated solutions   solutions that contain relatively large amounts of solute in relatively small amounts of solvent (or solution).

There are many methods of expressing concentrations of solutions. One method with which many are familiar is the proof rating of alcoholic beverages. The "proof" of an alcoholic beverage is twice the percent of alcohol in the beverage **by volume**. The karat rating of gold alloys (jewelry metal) indicates the percent of gold.

% Alcohol by volume	Proof		% Au by mass	Karat rating
_____	_____		_____	_____
_____	_____		_____	_____
_____	_____		_____	_____

Percent by Mass

Concentrations of solutions are often expressed as percent by mass of solute. This is the mass of solute per 100 mass units of solution. Symbolically, percent by mass of solute is

Example 3-11:   What mass of sodium hydroxide, NaOH, is required to prepare 250.0 g of solution that is 8.00% w/w NaOH by mass?

Example 3-12:   Calculate the mass of 8.00% w/w NaOH solution that contains 32.0 g of NaOH.

38

When the concentration of a solution is expressed as percent of solute, percent by mass is understood, unless percent by volume is specified. Also the solvent is understood to be water unless otherwise specified. The symbol for percent by mass is w/w. The symbol for percent by volume is v/v.

Example 3-13:   Calculate the mass of sodium hydroxide in 300.0 mL of solution that is 8.00% w/w NaOH. The density of the solution is 1.09 g/mL.

We have demonstrated that we can construct several unit factors from the description of 8.00% w/w sodium hydroxide solution. This information, and the knowledge that the density of the solution is 1.09 g/mL, enables us to construct the following unit factors. The reciprocal of each is also a unit factor.

Example 3-14:   What volume of 12.0% w/w potassium hydroxide, KOH, solution contains 40.0 grams of KOH? The density of the solution is 1.11 g/mL.

Molarity (Molar Concentration)

A common and very useful method of expressing concentrations of solutions is in terms of molarity (M). The *molarity* of a solution is the number of moles of solute per liter of solution. Symbolically, molarity is represented as

Thus, a 1 molar solution contains 1 mole of solute in enough water to make exactly one liter of solution. A 2 molar solution contains 2 moles of solute in enough water to make one liter of solution.

Frequently, it is convenient to express volumes of solutions in milliliters rather than liters, and to express the amount of solute in millimoles (mmol) rather than moles. Molarity may also be represented as

Example 3-15: Calculate the molarity of a solution that contains 12.5 grams of sulfuric acid, H_2SO_4, in 1.75 liters of solution.

Example 3-16: Calculate the mass of calcium nitrate, $Ca(NO_3)_2$, required to prepare 3.50 liters of 0.800 M $Ca(NO_3)_2$ solution.

Note in the above example that the volume of the solution multiplied by its molarity gives the number of moles of solute in the solution.

Concentrated solutions of commercial acids are used to prepare dilute solutions for laboratory use. Labels usually give specific gravity and the percent by mass of acid in a solution. It is necessary to calculate molar concentrations from these data.

Example 3-17: The specific gravity of commercial hydrochloric acid is 1.185 and it is 36.31 % w/w by HCl. What is its molarity.

40

3-7   Dilution of Solutions

In the dilution process a concentrated solution is added to water which produces a less concentrated, or more dilute, solution. In the dilution process the *number of moles of solute does not change*. However, molarity does change. Recall that multiplication of the volume of a solution (in liters) by its molarity (number of moles of solute/L) gives the number of moles of solute in the solution. Likewise the volume of a solution (in mL) can be multiplied by its molarity (number of mmol of solute/mL) to give the number of mmol of solute.

Example 3-18:   If 10. mL of 12 *M* hydrochloric acid is added to enough water to give 100. mL of solution, we have a dilute solution of HCl. Knowing the amount of concentrated HCl used, and the volume of new solution, we can calculate the concentration of the new solution.

Example 3-19:   What volume of 18.0 *M* sulfuric acid is required to prepare 2.50 liters of 2.40 *M* H_2SO_4 solution?

3-8   Using Solutions in Chemical Reactions

Example 3-20:   What volume of 0.500 *M* barium chloride, $BaCl_2$, solution is required for complete reaction with 4.32 grams of sodium sulfate, Na_2SO_4?

$$Na_2SO_4 \ + \ BaCl_2 \ \xrightarrow{\Delta} \ BaSO_4 \text{ (s)} \ + \ 2\,NaCl$$

41

Example 3-21:   (a) What volume of 0.200 molar sodium hydroxide, NaOH, solution will react
with 50.0 mL of 0.200 molar aluminum nitrate, $Al(NO_3)_3$, solution?

$$Al(NO_3)_3 \quad + \quad 3\,NaOH \quad \xrightarrow{\Delta} \quad Al(OH)_3\,(s) \quad + \quad 3\,NaNO_3$$

(b) What mass of aluminum hydroxide precipitates in (a)?

Titration is the process in which a solution of one reactant, the titrant, is carefully added to a solution of another reactant, and the volume of titrant required for complete reaction is measured.

Example 3-22:   What is the molarity of a potassium hydroxide solution if 38.7 mL of
the KOH solution is required to react with 43.2 mL of 0.223 M HCl?

$$KOH \quad + \quad HCl \quad \xrightarrow{\Delta} \quad KCl \quad + \quad H_2O$$

Example 3-23:   What is the molarity of a barium hydroxide solution if 44.1 mL of
0.103 M hydrochloric acid is required to react with 38.3 mL of the
$Ba(OH)_2$ solution?

$$Ba(OH)_2 \quad + \quad 2HCl \quad \xrightarrow{\Delta} \quad BaCl_2 \quad + \quad 2H_2O$$

Synthesis Question

Nylon is made by the reaction of hexamethylenediammine

$$H_2N-CH_2-C H_2-CH_2-CH_2-CH_2-CH_2-NH_2$$

with adipic acid

$$HOOC-CH_2-CH_2-CH_2-CH_2-COOH$$

in a 1 to 1 mole ratio. The structure of nylon is:

$$*-[-CO-CH_2-CH_2-CH_2-CH_2-CO-NH-CH_2-CH_2-CH_2-CH_2-CH_2-CH_2-NH-]_n-*$$

where the value of n is typically 450,000. On a daily basis, a DuPont factory makes 1.5 million pounds of nylon. How many pounds of hexamethylenediamine and adipic acid must be available in the plant each day?

Group Activity

Manganese(IV) oxide, potassium hydroxide, and oxygen react as indicated below.

$$4MnO_2 \ + \ 4KOH \ + \ 3O_2 \ \xrightarrow{\Delta} \ 4KMnO_4 \ + \ 2H_2O$$

A mixture of 272.9 g of MnO_2, 26.6 L of 2.50 M KOH, and 41.92 g of O_2 is allowed to react as shown above. When the reaction is complete, 234.6 g of $KMnO_4$ is separated from the reaction mixture. What is the per cent yield of this reaction?

Chapter Four

THE STRUCTURE OF ATOMS

SUBATOMIC PARTICLES

4-1   Fundamental Particles

Three fundamental particles make up atoms. The following table lists these particles together with their masses and their charges. The masses of the three fundamental particles are given in atomic mass units (amu), a term we shall describe shortly.

Fundamental Particles in Atoms

Particles	Symbol	Mass (amu)	Charge
electron	e⁻	0.00054858	-1
proton	p, p⁺	1.0073	+1
neutron	n, nº	1.0087	0

The experiments that provided conclusive evidence for the existence of the three fundamental particles give us insight into how we have learned about the structure of atoms.

4-2   The Discovery of Electrons

Davy (early 1800's) -

Faraday (1832-1833) -

47

Cathode ray tube experiments (Figure 4-1).

J. J. Thomson (1897)   - determined e/m ratio by measuring the degree of deflections of cathode rays by electrical fields of different magnitudes.

Millikan (1909) - determined the charge, then the mass of the electron (Figure 4-2).

4-3   Canal Rays and Protons

Goldstein (1886) - observed "Canal rays" (Figure 4-3).

4-4   Rutherford and the Nuclear Atom

In 1910, Geiger and Marsden bombarded metal foils with α-particles (positively charged particles; helium nuclei) from natural radioactive sources. See Figures 4-4 and 4-5.

Rutherford's analysis of the scattering of α-particles from the metal foils indicated that there were very small, very dense, positively charged centers in atoms. These are called atomic nuclei. Nuclei are surrounded by clouds of electrons at relatively large distances from the nuclei.

4-5   Atomic Number

A few years after Rutherford's experiments, Moseley studied x-rays given off by various elements (Figure 4-6). He correlated the wavelengths of emitted x-rays with atomic numbers of the target elements. See Figure 4-6. Now the known elements could be arranged in order of increasing nuclear charge (*atomic number*). Moseley's experiments provided evidence for increasing numbers of protons in the nuclei of atoms of increasing atomic weight. Elements are arranged in the periodic table in order of increasing atomic number. Hydrogen (Z=1) is the first element, helium (Z=2) is the second element, and so

A typical plot of some of Moseley's data (Figure 4-6) looks like

The *atomic number* (Z) of an element is the number of protons in the nuclei of its atoms. Atoms are electrically neutral; nuclei are positively charged (by the protons in the nucleus); beyond the nuclear radius there are a number of electrons equal to the number of protons in the nucleus. Therefore, the *atomic number* (Z) gives the number of protons in the nucleus and the number of electrons in the extranuclear structure of a neutral atom.

4-6   Neutrons

Neutrons were discovered by Chadwick in 1932 and are uncharged particles.

4-7   Mass Number and Isotopes

Isotopes are atoms of the same element that contain different numbers of neutrons in their nuclei. Or, isotopes may be defined as atoms of the same atomic number with different masses.

The *mass number* (A) is the number of protons plus the number of neutrons in the nucleus of an atom. Neutrons are electrically neutral; they contribute to the mass of an atom, but not to its charge. *Nuclide symbols* show both Z and A for atoms.

49

There are three isotopes of hydrogen.

$$_1^1 H \qquad\qquad _1^2 H \qquad\qquad _1^3 H$$

There are three naturally occurring isotopes of oxygen.

$$_8^{16} O \qquad\qquad _8^{17} O \qquad\qquad _8^{18} O$$

4-8 Mass Spectrometry and Isotopic Abundance

Isotopes have different masses. Their masses must be determined experimentally. Francis Aston built the first device to measure the masses of isotopes. A mass spectrometer is used to determine atomic masses. See Figures 4-8, 4-9, and 4-10.

Mass spectrometers measure the charge-to-mass ratio of charged particles (Figure 4-8). A sample of gas at very low pressure is bombarded with very high energy electrons. This causes electrons to be ejected from some of the gas molecules, forming positive ions. The positive ions are focused into a very narrow beam and accelerated by an electric field toward a magnetic field. The magnetic field deflects the ions from their straight-line path. The extent to which the beam of ions is deflected depends upon four factors:

1.

2.

3.

4.

A mass spectrometer is first calibrated using a sample of carbon-12. After the field strength required to focus the beam of carbon-12 ions has been measured, a sample of another element is injected into the mass spectrometer. The field strength required to focus this beam of ions is measured.

The mass spectrometer is used to measure the masses of isotopes as well as isotopic abundances. See Figure 4-9, the mass spectrum of Ne^+ ions.

50

4-9   The Atomic Weight Scale and Atomic Weights

The mass of the carbon-12 atom, $^{12}_{6}C$, is defined as exactly 12 atomic mass units (amu). One atomic mass unit is exactly 1/12th the mass of a carbon-12 atom.

Example 4-1:   Calculate the number of atomic mass units in one gram.

In Chapter 2, we stated a working definition of atomic weight. We now define the term precisely. *The atomic weight of an element is the weighted average of the masses of its constituent isotopes.*

Example 4-2:   Naturally occurring copper consists of two isotopes. It is 69.1% $^{63}_{29}Cu$, which has a mass of 62.9 amu, and 30.9% $^{65}_{29}Cu$, which has a mass of 64.9 amu. Calculate the atomic weight of Cu to one decimal place.

Example 4-3:   Naturally occurring chromium consists of four isotopes. It is 4.31% $^{50}_{24}Cr$, mass = 49.946 amu, 83.76% $^{52}_{24}Cr$, mass = 51.941 amu, 9.55% $^{53}_{24}Cr$, mass = 52.941 amu, and 2.38% $^{54}_{24}Cr$, mass = 53.939 amu. Calculate the atomic weight of chromium.

If there are only two naturally occurring isotopes of an element, and their masses are known, then the fraction of each can be calculated from the atomic weight of the element.

Example 4-4:   The atomic weight of boron is 10.811 amu. The masses of the two naturally occurring isotopes $^{10}_{5}B$ and $^{11}_{5}B$, are 10.013 and 11.009 amu, respectively. Calculate the fraction and percentage of each isotope.

4-10 The Periodic Table: Metals, Nonmetals, and Metalloids

1869 - Mendeleev published a periodic table similar to the current table based *primarily on chemical properties* of known elements. See Figure 4-1.

1869 - Meyer published a similar table based *primarily on physical properties.* Both tables emphasized *periodicity*, the regular variations of properties with increasing *atomic weight*. The concept of *atomic number* had not been developed in 1869.

Mendeleev noted that both chemical and physical properties of elements vary in a periodic (repeating) pattern, a summary of observations that is known as the periodic law. The law is now stated: *The properties of the elements are periodic functions of their atomic numbers.* In 1871 he predicted some properties of the (then) unknown element germanium (number 32) and several of its compounds (Table 4-1).

Let us review some of the terminology we use to describe the periodic table.

Group (family) of elements

Period (row) of elements

Group IA metals

Group 2A metals

Group 6A nonmetals

Group 7A elements

Metals, Nonmetals, and Metalloids

In a very broad and general way, elements may be classified as *metals* and as *nonmetals. Except for hydrogen*, elements to the left of the heavy stair step on the periodic table are *metals.* Those to the right are *nonmetals*. Elements adjacent to this line are called *metalloids* because they have some properties of both metals and nonmetals. See Tables 4-2, 4-3, and 4-4.

| Trends in metallic charater |

Some properties of metals and nonmetals are tabulated in Tables 4-3 and 4-4. Study these tables carefully.

THE ELECTRONIC STRUCTURES OF ATOMS

4-11 Electromagnetic Radiation

Electromagnetic radiation (radiant energy) is characterized by its *wavelength (color)*, λ, (lambda) and its frequency (*energy*), ν, (nu). They are related by the equation, $\lambda\nu = c$, where c is the speed of light through a vacuum, 3.00×10^8 m/s. *Wavelength* refers to the distance between successive crests or troughs. *Frequency* describes the number of crests or (troughs) passing a given point per second. The *amplitude* refers to the intensity of the wave. See Figure 4-11.

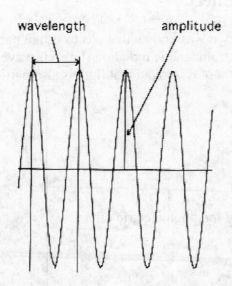

Example 4-5: What is the frequency of green light of wavelength 5200 Å?

Pay attention to the various types of electromagnetic radiation and their relative energies and frequencies shown in Figure 4-13.

Light can also be described in terms of very small particles, called *photons*. In 1900 Max Planck suggested that each photon of light has a definite amount (a *quantum*) of energy. The energy of a photon of light is

where:
h = Planck's constant, 6.63×10^{-34} J•s
ν = frequency of light, s^{-1}
λ = wavelength

53

We see that the energy of light is directly proportional to its frequency, ν, and inversely proportional to its wavelength, λ.

Example 4-6:   What is the energy of a photon of green light of wavelength 5200 Å? From Example 4-5, we know that $\nu = 5.77 \times 10^{14}$ s^{-1}.

4-12 The Photoelectric Effect

Early in the 20th century it was known that electrons are particle-like (possess definite mass and charge), and that electromagnetic radiation (light) is wave-like. The photoelectric effect, and Einstein's explanation of it, proved that light is also particle-like, as Planck had suggested. See Figure 4-14.

What are some practical uses of the photoelectric effect?

4-13 Atomic Spectra and the Bohr Atom

Passing an electric current through a gas in a vacuum tube (at very low pressure) causes the gas to emit light. This light can be dispersed by a prism to give an *emission spectrum* or a *bright line spectrum*. See Figure 4-15a. These lines can be recorded photographically, and the wavelength of the light that produced each line can be calculated from its position on the film.

Shining a beam of white light (continuous distribution of wavelengths) through a sample of gas gives an *absorption spectrum*, which shows the wavelengths of light that have been *absorbed*. See Figure 4-15b.

For a given gas, the wavelengths of light absorbed in the absorption spectrum, are the same wavelengths that are emitted in the emission spectrum. (Figure 4-15). Each element displays its own characteristic set of lines (Figure 4-16). These spectra are "fingerprints" that allow us to identify each element.

Example 4-7:   An orange line of wavelength 5890 Å is observed in the emission spectrum of sodium. What is the energy of one photon of this orange light?

Example 4-8:   What is the wavelength of light emitted when the hydrogen's atom energy changes from n = 4 to n = 2?

Several series of lines in the spectrum of hydrogen are produced by passing electric current through the gas at very low pressures. See Figures 4-16. Late in the 19th century J. R. Rydberg discovered that the wavelengths of the lines in the hydrogen spectrum could be related by an empirical mathematical equation.

$$R = 1.097 \times 10^7 \text{ m}^{-1}$$
(the Rydberg constant)

$$n_1 < n_2$$

The greatest successes of the quantum theory have been in the interpretation of spectra of atoms, ions, and molecules. Older theories of atomic structure were *unable* to explain the observed spectra of simple species, i.e., the spectrum of atomic hydrogen. The Rutherford model of the atom was that of a nuclear system. Although it fairly satisfactorily explained chemical reactions in its day, it was not capable of providing an explanation for, or an interpretation of, simple atomic spectra.

Bohr's theory suggested that spectral lines were the result of electrons moving between energy levels in the atom. When an electron moves from a *lower* energy level to a *higher* energy level in an atom ($n_1 \rightarrow n_2$), energy of a characteristic frequency (wavelength) is *absorbed*. When an electron falls from the second energy level back to the first ($n_2 \rightarrow n_1$), then radiation of the same frequency (wavelength) is *emitted*.

4-14 The Wave Nature of the Electron

de Broglie predicted (1925) that electrons are also wave-like, with wavelengths described by the equation

h = Planck's constant
m = mass of electron
ν = velocity of electron

In only two years, de Broglie's predictions were verified at the Bell Telephone Laboratory. The wave nature of electrons had been established by 1928! This property is exploited by the electron microscope.

4-15 The Quantum Mechanical Picture of the Atom

We now know that electrons can be treated more effectively as waves than as small particles.

The Heisenberg Uncertainty Principle points out a weakness of the Bohr Theory. This principle is one of the fundamental ideas of quantum mechanics. *It is impossible to determine simultaneously both the momentum and the position of an electron (or any other very small particle).* The momentum of a particle is the product of its mass and velocity, i.e., mv. (Angular momentum is mvr). Because the exact position or path of an electron cannot be determined, we speak of the *probability* of finding an electron within specified regions within an atom. There are definite energy levels that electrons can occupy in atoms. However, it is not possible to determine precisely where an electron is.

The basic ideas of quantum mechanics are:

1.   Atoms and molecules can exist only in certain energy states. In each energy state, the atom or molecule has a definite energy. When an atom or molecule changes its energy state, it must emit or absorb just enough energy to bring it to the new energy state (the quantum condition).

Atoms and molecules possess various forms of energy. Let us focus our attention on their electronic energies.

2.   Atoms or molecules emit or absorb radiation (light) as they change their energies. The frequency of the light emitted or absorbed is related to the energy change by a simple equation.

This gives a relationship between the energy change ΔE, and the wavelength, λ, of the radiation emitted or absorbed. The energy lost (or gained) by an atom as it goes from higher to lower (or lower to higher) energy states is equal to the energy of the photon emitted (or absorbed) during the transition.

3.   The allowed energy states of atoms and molecules can be described by sets of numbers called *quantum numbers*.

Because we cannot know both the position (or path) and momentum (or energy) of an electron simultaneously, we resort to a statistical approach to describe and predict electronic configurations. The solution of the Schrodinger equation, as modified by Dirac, is the basis for this treatment. It produces a set of *four quantum numbers* for each electron in an atom.

4-16 Quantum Numbers

Quantum numbers are very important in describing the energy levels in atoms and the shapes of the atomic orbitals that describe distributions of electrons in space. **An atomic orbital is a region in space in which the probability of finding an electron is high.**

The **four quantum numbers** distinguish among each and every electron in an atom.

1.   Principal Quantum Number, n.   The principal quantum number indicates the effective volume in space in which the electron moves about the nucleus. The principal quantum number, n, corresponds to a main energy level. The allowed values of n are

2.   Subsidiary Quantum Number, ℓ. The subsidiary quantum number designates the shape of the region (volume) that an electron occupies. Subsidiary quantum numbers refer to the kinds of atomic orbitals, i.e., *s, p, d, f,* ...orbitals. Theoretically, this range may be extended to g, h, i,...etc. orbitals. The range of values that ℓ may take is determined by n.

3.   Magnetic Quantum Number, $m\ell$. The magnetic quantum number indicates the spatial orientation of an atomic orbital. The range of values that $m\ell$ may take is determined by the value of ℓ.

4.   Spin Quantum Number, m_S. The spin quantum numbers refers to the spin of the electron and the orientation of the magnetic field produced by this spin. The values of m_S are either $+1/2$ or $-1/2$.

4-17 Atomic Orbitals

The regions about the nucleus in which the probability of finding electrons is greatest are referred to as *atomic orbitals*. There are several kinds of atomic orbitals. Energy levels may be described in terms of the principal quantum number, *n*, or by the letter designations, K, L, M, N, O, etc.

s Atomic Orbitals

s orbitals are spherically symmetrical with respect to the nucleus. There is one s orbital per energy level. The quantum number $\ell = 0$ describes s atomic orbitals. The quantum number $m\ell$ can only be 0 when $\ell = 0$. See Figures 4-21 and 4-22.

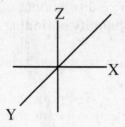

p Atomic Orbitals

A set of p orbitals consists of three mutually perpendicular equal-arm dumbbells. There are three p orbitals per energy level beginning with the second energy level. The lobes of p orbitals are directed along the axes of a set of Cartesian coordinates. The quantum number $\ell = 1$ describes atomic orbitals. When $\ell = 1$, $m\ell$ can take the values -1, 0, +1. These values describe the three p orbitals in an energy level. See Figures 4-22 and 4-23.

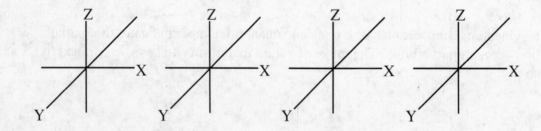

58

d Atomic Orbitals

There are five *d* orbitals per energy level beginning with the third energy level. The quantum number $\ell = 2$ describes *d* atomic orbitals. When $\ell = 2$, m_ℓ can take the values -2, -1, 0, +1, +2. These describe the five *d* orbitals in an energy level. See Figure 4-24.

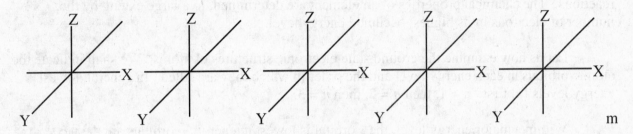

f Atomic Orbitals

There are seven *f* atomic orbitals per energy level *beginning with the fourth energy level.* The quantum number $\ell = 3$ describes *f* atomic orbitals. When $\ell = 3$, m_ℓ can take the values -3, -2, -1, 0, +1, +2, +3. See Figure 4-25. The *f* orbitals are the last orbitals occupied by electrons in presently known atoms in their ground states. There are higher energy orbitals in the fifth and higher energy levels, but they are occupied only in excited states.

Each atomic orbital can accommodate two electrons. For two electrons to occupy the same orbital, their spins must be paired. The term *spin*, as applied to an electron, refers to the magnetic field generated by the motion of an electron. The motion of an electric charge of necessity generates a magnetic field. Because electrons have electrical charges, their motion generates magnetic fields. For two electrons to occupy an atomic orbital, their magnetic fields *must* interact attractively, i.e., their spins must be "paired".

The number of atomic orbitals per (major) energy level is n^2, where *n* represents the principal quantum number, i.e., the major energy level; the maximum number of electrons that can occupy a major energy level is $2n^2$.

Energy Level n	Number of Orbitals n^2	Maximum Number of Electrons $2n^2$
1	_____	_____
2	_____	_____
3	_____	_____
4	_____	_____

4-18 Electron Configurations

As we study electronic structures of atoms, we keep in mind the *Periodic Law. The properties of the elements are periodic functions of their atomic numbers.* We also keep in mind that electrons in the highest occupied energy levels of atoms are the ones involved in chemical reactions. The chemical properties of an element are determined, to a large extent, by the number of electrons in its highest occupied energy level.

Let us now examine the ground state electronic structures of atoms. We shall indicate the atomic orbitals in each energy level and the order in which they are filled. In general, lowest energy levels fill first: $n = 1$, then $n = 2$, then $n = 3$, …

Within a major energy level, the s orbital is lowest in energy, p orbitals are the next lowest, then d orbitals, and finally f orbitals are highest in energy.

There are several important rules that determine the order in which the electrons fill ground state atomic energy levels. These include

the *Aufbau Principle* , **The electron that distinguishes an element from the previous element enters the lowest-energy atomic orbital available.**

the *Pauli Exclusion Principle*, **No two electrons in an atom may have identical sets of four quantum numbers.**

and *Hund's rule*, **Electrons must occupy all of the orbitals in a given sublevel singly before pairing begins. These unpaired electrons have parallel spins.**

There are two elements in the first period (row) of the periodic table, hydrogen and helium. The electrons in hydrogen and helium occupy the $1s$ orbital.

	$1s$	Outermost configuration
$_1$H	___	_____
$_2$He	___	_____

60

There are eight elements (numbers 3-10) in the second row of the periodic table. The outermost electrons in these elements occupy the 2s and 2p orbitals; that is, they all have $n = 2$, where n is the *outermost occupied energy level*.

	1s	2s	2p	Outermost Configuration
$_3$Li	__	__	__ __ __	_____
$_4$Be	__	__	__ __ __	_____
$_5$B	__	__	__ __ __	_____
$_6$C	__	__	__ __ __	_____
$_7$N	__	__	__ __ __	_____
$_8$O	__	__	__ __ __	_____
$_9$F	__	__	__ __ __	_____
$_{10}$Ne	__	__	__ __ __	_____

There are also eight elements (numbers 11-18) in the third period. The outermost electrons in these elements occupy the 3s and 3p orbitals ($n = 3$). The symbol for a noble gas indicates the number of electrons, corresponding to the atomic number of the noble gas, that are in filled sets of orbitals. For example, [Ne] corresponds to 10 e⁻.

		3s	3p	Outermost Configuration
$_{11}$Na	[Ne]	__	__ __ __	_____
$_{12}$Mg	[Ne]	__	__ __ __	_____
$_{13}$Al	[Ne]	__	__ __ __	_____
$_{14}$Si	[Ne]	__	__ __ __	_____
$_{15}$P	[Ne]	__	__ __ __	_____
$_{16}$S	[Ne]	__	__ __ __	_____
$_{17}$Cl	[Ne]	__	__ __ __	_____
$_{18}$Ar	[Ne]	__	__ __ __	_____

There are eighteen elements (numbers 19-36) in the fourth period of the periodic table. As we shall see, the electrons added in these elements occupy the 4s, 3d, and 4p orbitals. [Ar] corresponds to 18 e⁻ in filled sets of orbitals.

		3d	4s	4p	Outermost Configuration
19K	[Ar]	__ __ __ __ __	__	__ __ __	_____
20Ca	[Ar]	__ __ __ __ __	__	__ __ __	_____
21Sc	[Ar]	__ __ __ __ __	__	__ __ __	_____
22Ti	[Ar]	__ __ __ __ __	__	__ __ __	_____
23V	[Ar]	__ __ __ __ __	__	__ __ __	_____
24Cr	[Ar]	__ __ __ __ __	__	__ __ __	_____
25Mn	[Ar]	__ __ __ __ __	__	__ __ __	_____
26Fe	[Ar]	__ __ __ __ __	__	__ __ __	_____
27Co	[Ar]	__ __ __ __ __	__	__ __ __	_____
28Ni	[Ar]	__ __ __ __ __	__	__ __ __	_____
29Cu	[Ar]	__ __ __ __ __	__	__ __ __	_____
30Zn	[Ar]	__ __ __ __ __	__	__ __ __	_____
31Ga	[Ar]	__ __ __ __ __	__	__ __ __	_____
32Ge	[Ar]	__ __ __ __ __	__	__ __ __	_____
33As	[Ar]	__ __ __ __ __	__	__ __ __	_____
34Se	[Ar]	__ __ __ __ __	__	__ __ __	_____
35Br	[Ar]	__ __ __ __ __	__	__ __ __	_____
36Kr	[Ar]	__ __ __ __ __	__	__ __ __	_____

Let us write an acceptable set of quantum numbers for the electrons in a few elements.

$$n \qquad \ell \qquad m_\ell \qquad m_s$$

$_{11}$Na

$_{20}$Ca   We use the symbol [Ar] to represent the first 18 electrons in Ca.

$_{26}$Fe   We use the symbol [Ar] to represent the first 18 electrons in Fe.

As you study the rest of this chapter, you will see that the figures and tables give you lots of very useful information!

4-19   The Periodic Table and Electron Configurations

The *vertical divisions* in the periodic table are called *groups* or *families* of elements. The *horizontal divisions* are called *rows* or *periods* of elements.

The pattern for filling atomic orbitals has been evolved. Study Table 4-5 carefully. There are several mnemonic devices to assist one in remembering the order in which electrons are "added" to atoms. See Figure 4-32. The best mnemonic for determining the electron configurations of elements is the Periodic Table. It can be broken into blocks that represent the various electron orbitals that are being filled.

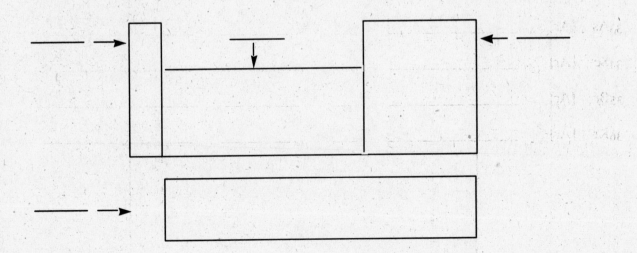

4-20 Paramagnetism and Diamagnetism

Unpaired electrons have their spins aligned _____ or _____. Atoms with unpaired electrons are called **paramagnetic**.

Paired electrons have their spins unaligned _____.

Atoms with unpaired electrons are called **diamagnetic**.

Synthesis Question

What is the atomic number of the element that should theoretically be the noble gas below Rn?

Group Question

In a universe different from ours, the laws of quantum mechanics are the same as ours with one small change. Electrons in this universe have three spin states, -1, 0, and +1, rather than the two, +1/2 and -1/2, that we have. Which two elements in this universe would be the first and second noble gases? (Assume that the elements in this different universe have the same symbols as in ours.)

Chapter Five

CHEMICAL PERIODICITY

In this chapter we shall correlate electronic configurations of elements with their positions in the periodic table and with their properties. Many properties of the A Group elements and the noble gases vary regularly with their positions in the periodic table. Variations among the B Group elements are not nearly so regular, and so we shall focus our attention on A Group elements.

5-1   More About The Periodic Table

The Noble Gases

With the exception of helium, the noble gases have eight electrons in their highest occupied energy level. They occupy Group 8A in the periodic table.

Representative Elements

The A Group elements in the periodic table are referred to as representative elements. These are elements in which electrons are being added to either *s* or *p* orbitals.

d-Transition Elements

The B Group elements are known as *d*-transition elements. All are metals. These are elements in which electrons are being added to *d* orbitals. Because the *d* orbitals that are being filled are one shell *inside* the outermost occupied shell, variations in properties among the *d*-transition elements are not nearly as great (across a row) as among the representative elements.

f-Transition Elements

Elements 58-71 and 90-103 are known as inner-transition metals or *f*-transition metals. These are elements in which electrons are being added to *f* atomic orbitals two shells inside the outermost occupied shell. Variations in properties of these elements are very slight indeed, as we move across a given row.

The outermost electrons have the greatest influence on the properties of elements.

PERIODIC PROPERTIES OF THE ELEMENTS

Many properties of elements vary with position in the periodic table. Some properties that are important in predicting chemical properties and in understanding chemical bonding follow.

5-2   Atomic Radii (A Group Elements Only)

The parameter used to describe the sizes of atoms is their atomic radius. Atoms are much too small to be measured individually. Radii of atoms are calculated from measurements on pure samples of solid elements. Atomic radii *decrease from left to right* in the periodic table and *increase from top to bottom*. See Figure 5-1. The decrease in radii from left to right across the periodic table is explained by the *shielding (screening) effect*. Electrons in *filled sets* of *s* and *p* orbitals (noble gas configurations) fairly effectively "screen out" the attractive effect on the *outermost* electrons of a number of protons equal to the number of electrons in the filled sets.

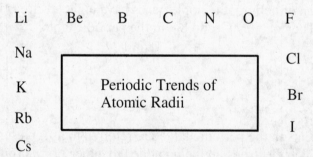

Example 5-1:   Arrange these elements based on their atomic radii.

Se, S, O, Te

Example 5-2:   Arrange these elements based on their atomic radii.

P, Cl, S, Si

Example 5-3:   Arrange these elements based on their atomic radii.

Ga, F, S, As

5-3   Ionization Energy

The first ionization energy (IE) of an element is the minimum amount of energy required to remove the most loosely held electron from an isolated gaseous atom. For example

The *second ionization energy* is the amount of energy required to remove the second electron.

Ionization energies measure how tenaciously atoms hold on to their electrons. The metals, which are located toward the left of the periodic table, have lower ionization energies than the nonmetals, which are located toward the right of the table. See Table 5-1 and Figure 5-2. We see that first ionization energies generally *increase* as we move *toward the right* and they generally *decrease* as we move *down a group* in the periodic table.

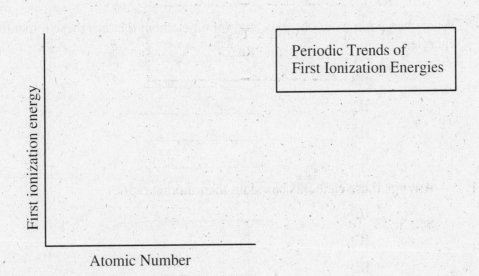

Periodic Trends of
First Ionization Energies

Example 5-4:   Arrange these elements based on their first ionization energies.

              Sr, Be, Ca, Mg

Example 5-5:   Arrange these elements based on their first ionization energies.

              Al, Cl, Na, P

Example 5-6:   Arrange these elements based on their first ionization energies.

              B, O, Be, N

Consider the first 4 ionization energies (in kJ/mol) of the following elements of period 3:

	IA	2A	3A	4A
Period 3	Na	Mg	Al	Si
1st IE	____	____	____	____
2nd IE	____	____	____	____
3rd IE	____	____	____	____
4th IE	____	____	____	____

The stable ions of these elements are Na^+, Mg^{2+}, and Al^{3+}. Silicon does not form simple monatomic cations. In general, the maximum charges on monatomic ions are 3^+ and 3^-.

Example 5-7:   What charge ion would be expected for an element that has these ionization energies.

              IE_1 _____

              IE_2 _____

              IE_3 _____

              IE_4 _____

              IE_5 _____

              IE_6 _____

              IE_7 _____

              IE_8 _____

5-4 Electron Affinity

The electron affinity (EA) of an element is the amount of energy absorbed when an electron is added to an isolated gaseous atom to form an anion with a 1- charge. EA's have negative values when energy is released and positive values when energy is absorbed.

Electron affinities usually become more negative from *left to right* and from *bottom to top* in the periodic table. There are many exceptions. See Figure 5-3 and Table 5-2.

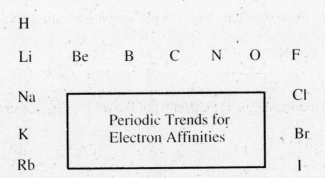

Example 5-8:   Arrange these elements based on their electron affinities.

Al, Mg, Si, Na

5-5 Ionic Radii

Simple *cations* are formed when atoms lose electrons and become positively charged ions. *Cations are always smaller than their parent neutral atoms.* See Figure 5-4.

Anions are formed when neutral atoms gain electrons to form negatively charged ions. *Anions are always larger than their parent neutral atoms.* See Figure 5-4.

71

N O F

N^{3-} O^{2-} F^-

Variations in Radii of Anions (A Groups Only)

Isoelectronic species have the same number of electrons. Notice carefully the variations in radii of isoelectronic atoms and ions as their nuclear charges change.

Example 5-9: Arrange these elements based on their ionic radii.

Ga, K, Ca

Example 5-10: Arrange these elements based on their ionic radii.

Cl, Se, Br, S

5-6 Electronegativity

Electronegativity (EN) is not as precise a term as electron affinity. However, electronegativity values are very useful. The electronegativity of an atom is a measure of the tendency of an atom to attract electrons to itself when that atom is in combination with another atom.

Electronegativities usually increase from left to right in the periodic table. They decrease from top to bottom. There are a few exceptions. See Table 5-3.

H

Li Be B C N O F

Na Cl

K Br

Rb I

Periodic Trends for Electronegativity

Example 5-11: Arrange these elements based on their electronegativity.

Se, Ge, Br, As

Example 5-12: Arrange these elements based on their electronegativity.

Be, Mg, Ca, Ba

We may summarize some of the trends examined to this point.

1.   **Elements (*metals*) with low ionization energies and low electronegativities lose electrons easily to form positively charged monatomic ions (cations).**

2.   **Elements (*nonmetals*) with high ionization energies, highly negative electron affinities, and high electronegativities gain electrons readily to form negatively charged monatomic ions (anions).**

3.   **Simple *cations* are always *smaller* than their parent atoms.**

4.   **Simple *anions* are always *larger* than their parent atoms.**

5-7   Oxidation Numbers

Reactions in which substances undergo changes in oxidation number are called *oxidation-reduction reactions*, or simply *redox reactions*. Oxidation and reduction always occur simultaneously and to the same extent in chemical reactions. Table 4-10 lists common oxidation numbers.

Oxidation number/state

-   the number of electrons gained (-) or lost (+) by an atom when it forms a simple ionic compound
-   in covalent compounds the less electronegative element is assigned a positive oxidation number and the more electronegative element is assigned a negative oxidation number, according to the following guidelines.

Oxidation Number Guidelines

1.   The oxidation number of any free, uncombined element is zero.

2.   The oxidation number of an element in a simple (monatomic) ion is the charge on the ion.

3.   In the formula for *any* compound, the sum of the oxidation numbers of all elements in the compound is zero.

4.   In a polyatomic ion, the sum of the oxidation numbers of the constituent elements is equal to the charge on the ion.

5.   Flurine has an oxidation number of -1 in its compounds.

6.   Hydrogen, H, usually has the oxidation number +1 in its compounds.

7.   Oxygen, O, usually has the oxidation number -2 in its compounds.

Elements	Common Oxidation Numbers
1. Group IA metals	+1
2. Group 2A metals	+2
3. Group 3A metals	+3
4. Group 7A elements in binary compounds with metals	-1
5. Group 6A elements in binary compounds with metals	-2
6. Group 5A elements in binary compounds with metals	-3

Example 5-13:   Assign oxidation numbers to each element in the following:

$NaNO_3$
$K_2Sn(OH)_6$
H_3PO_4

$SO_3{}^{2-}$
$HCO_3{}^{-}$
$Cr_2O_7{}^{2-}$

CHEMICAL REACTIONS AND PERIODICITY

5-8   Hydrogen and the Hydrides

Hydrogen

Hydrogen, H_2, is a colorless, odorless, tasteless, flammable diatomic gas. It can be prepared in the laboratory by the reaction of an active metal, such as magnesium, with a nonoxidizing acid, such as hydrochloric acid.

Combustion is the highly exothermic combination of a substance with oxygen, usually with a flame. The combustion of H_2 liberates a great amount of heat.

74

Hydrogen is prepared industrially by several methods. One method is the *thermal cracking of hydrocarbons*. This involves heating a hydrocarbon in the *presence* of a catalyst and in the *absence* of air.

Reactions of Hydrogen and the Hydrides

Hydrogen reacts with very active *metals* to produce solid **ionic** compounds, called ionic hydrides that contain hydride ions, H^-. Consider the reaction of H_2 with molten potassium (IA).

In general, for the IA metals

Molten barium (Group 2A) reacts with H_2 to produce barium hydride.

In general, for the 2A metals

The ionic hydrides are basic; H^- ions reduce water to produce H_2 and OH^- ions.

Hydrogen reacts with *nonmetals* to produce molecular hydrides, many of which are acidic. H_2 reacts with the halogens (Group 7A) to form covalent hydrogen halides; all are gases at normal temperatures and pressures.

$$X = F, \ Cl, \ Br, \ I$$

Specifically, the reactions of hydrogen with fluorine and bromine are

Hydrogen reacts with oxygen and other elements of Group 6A to produce the covalent compounds H_2O, H_2S, H_2Se, and H_2Te.

75

The ionic or molecular hydrides of representative elements are displayed in Figure 5-5.

One important trend that can be discerned is

Ionic hydrides are all basic, whereas molecular (nonmetal) hydrides are all acidic.

5-9   Oxygen and the Oxides

Oxygen and Ozone

Oxygen, O_2, was discovered in 1774 by Priestly as a product of the thermal decomposition of mercury(II) oxide.

Commercially, O_2 is obtained by fractional distillation of liquefied air. O_2 can be prepared by a number of methods in the laboratory, such as the thermal decomposition of potassium chlorate in the presence of a catalyst, MnO_2.

Oxygen also exists in another **allotropic form**, ozone, O_3, which is a very strong oxidizing agent. The O_3 molecule is angular and diamagnetic.

O_3 is formed from O_2 in the upper atmosphere (and elsewhere) when O_2 molecules absorb certain wavelengths of ultraviolet radiation.

Reactions of Oxygen and the Oxides

Oxygen forms oxides by combination with all other elements except for the noble gases and unreactive metals (Au, Pd, Pt). An oxide is a binary compound that contains oxygen.

Reactions of O_2 with Metals

Oxygen reacts with all but the least active metals (Pt, Pd, Au, the so-called noble metals) to produce primarily solid *ionic metal oxides*. The major products of reactions of the Group IA metals with O_2 differ for different metals. Lithium reacts with O_2 to produce lithium oxide, a *normal* oxide which contains oxide ions, O^{2-}.

Sodium reacts with O_2 to form mostly sodium peroxide. The oxidation number of oxygen is -1 in peroxide ions, O_2^{2-}.

Reactions of O_2 with K, Rb, and Cs produce mainly superoxides. The superoxide ion is, O_2^{1-}, in which the oxidation number of oxygen is -1/2.

Oxygen reacts with Group 2A metals to form normal oxides. In general,

Specifically, the reaction of strontium with O_2 at atmospheric pressure is

Under high pressures of O_2 the heavier members, Ca, Sr, and Ba, also form peroxides.

Metal oxides are formed when less active metals are heated in O_2. For metals that exhibit variable oxidation states, the oxide formed depends upon the relative amounts of metal and oxygen used. Consider the reactions of manganese.

Limited O_2

Excess O_2

Reactions of Metal Oxides with Water

Many *metal oxides (basic anhydrides)* react with water to form *ionic metal hydroxides, bases* with no change in oxidation state of the metal. See Figure 5-9.

Lithium oxide reacts with water to form lithium hydroxide

Calcium oxide reacts with water to form calcium hydroxide

Reactions of O_2 with Nonmetals

Oxygen reacts with *nonmetals* to form *molecular oxides*. Consider the reaction of hot carbon with a limited amount of O_2.

With excess O_2, the product is carbon dioxide.

The reaction of phosphorus with a limited amount of O_2 produces tetraphosphorus hexoxide.

If an excess of O_2 is used in the reaction tetraphosphorus decoxide is formed.

Reactions of Nonmetal Oxides with Water

Many *nonmetal oxides (acid anhydrides)* react with water to produce ternary acids. Many acid anhydrides dissolve in water to form acid with no change in oxidation state of the nonmetal.

Carbon dioxide reacts with water to form carbonic acid.

Dichlorine heptoxide reacts with water to form perchloric acid.

78

Diarsenic pentoxide reacts with hot water to form arsenic acid.

Reactions of Metal Oxides with Nonmetal Oxides

Nonmetal oxides (acid anhydrides) react with *metal oxides* (basic anhydrides) to form salts. A salt contains the cation of a base and the anion of an acid. For example, carbon dioxide reacts with lithium oxide to form lithium carbonate, a salt.

Dichlorine heptoxide reacts with magnesium oxide to form magnesium perchlorate, a salt.

Combustion Reactions

These are redox reactions in which oxygen reacts with oxidizable materials in highly exothermic reactions. Fossil fuels are primarily hydrocarbons. The *complete* combustion of hydrocarbons produces carbon dioxide and water. Examples include

Fossil fuels are always "contaminated" with sulfur, which produces the harmful air pollutant, sulfur dioxide during combustion.

SO_2 is slowly oxidized to SO_3 in the air.

SO_3 is the acid anhydride of sulfuric acid.

H_2SO_4 is the major contributor to acid rain.

Combustion of Fossil Fuels and Air Pollution

Nitrogen compounds are also impurities in fossil fuels. These undergo combustion to form nitrogen oxide. *Most* nitrogen oxide is produced from the air that is mixed with the fuel. The NO is exhausted into the atmosphere from furnaces, automobiles, and airplanes. NO is oxidized to NO_2 by oxygen, especially in the presence of ultraviolet light.

Nitrogen dioxide is the colored component of the reddish-brown haze of photochemical smog. NO_2 also reacts with moisture in air to form nitric acid (in a disproportionation reaction).

NO_2 can react with some organic compounds to form other pollutants.

Synthesis Question

When the elements Np and Pu were first discovered by McMillan and Seaborg, they were placed on the periodic chart just below La and Hf. However, after studying the chemistry of these new elements for a few years, Seaborg decided that they should be placed in a new row beneath the lanthanides. What justification could Seaborg have used to move these elements on the periodic chart?

Group Question

What do the catalytic converters that are attached to all of our cars' exhaust systems actually do? How do they decrease air pollution?

Chapter Six

SOME TYPES OF CHEMICAL REACTIONS

As chemical reactions occur, old bonds may be broken and new bonds formed, molecules and ions may lose or gain electrons, or insoluble combinations of ions may form precipitates. We shall consider some of the reactions that may occur. First, we consider some important elements and the periodicity of their properties.

6-1   Aqueous Solutions - An Introduction

Because many reactions occur in aqueous (water) solutions, we must know the kinds of substances that are soluble in water and the predominant forms in which dissolved substances exist in aqueous solutions or in contact with water.

1   Electrolytes and Extent of Ionization

Compounds that are water-soluble are classified as *electrolytes* or *nonelectrolytes*. Compounds whose aqueous solutions conduct electricity are called *electrolytes* because their solutions contain ions that conduct electricity (Figure 6-1). *Nonelectrolytes* do not conduct electricity because their solutions do not contain ions.

Examples of nonelectrolytes include:

Strong electrolytes are nearly completely ionized (or dissociated) in dilute aqueous solution. The common strong electrolytes are: (1) strong acids (Table 6-1), (2) strong soluble bases (Table 6-3), and (3) soluble ionic salts.

Weak electrolytes are only slightly ionized (or dissociated) in dilute aqueous solution. Weak acids (Tables 6-2), weak bases, and a few soluble covalent salts are weak electrolytes.

83

2   Strong and Weak Acids

Acids are compounds that produce hydrogen ions (H^+ or H_3O^+) in aqueous solutions. *Strong acids* ionize completely, or nearly so, in dilute aqueous solutions. Nitric acid is a typical strong acid.

The common strong acids follow. See Table 6-1 for their names.

_____   _____   _____   _____

_____   _____   _____   _____

_____   _____   _____   _____

_____   _____

Weak acids ionize only slightly in dilute aqueous solutions.

A few common weak acids follow. See Table 6-2 and Appendix F for their names.

_____   _____   _____   _____

_____   _____   _____   _____

_____   _____   _____   _____

The formulas for inorganic (mineral) acids are usually written with H first. Organic acids contain the carboxyl (-COOH) group. Nearly all organic acids are weak acids.

Mineral acids ionize as follows:

3   Reversible Reactions

A double arrow ($\rightleftharpoons$) indicates that the reaction is *reversible*, i.e., it occurs in both directions.

Weak electrolytes, such as acetic acid, ionize only slightly (and reversibly) in dilute aqueous solutions.

4   Strong Bases, Insoluble Bases, and Weak Bases

Bases are compounds that produce OH⁻ ions in aqueous solutions. The *strong soluble* bases are ionic, even in the solid state. They *dissociate* completely in dilute aqueous solution.

For potassium hydroxide we represent the dissociation as

For barium hydroxide the dissociation is represented as

The strong soluble bases are the hydroxides of Group IA metals and the heavier members of Group 2A. See Table 6-3 for their names.

_____   _____

_____   _____

_____   _____   _____   _____

_____   _____   _____   _____

_____   _____   _____   _____

Many other metal hydroxides are ionic but they are so insoluble that they cannot produce strongly basic solutions; these are called *insoluble bases*. Examples include

The weak bases are covalent substances that are soluble in, but that ionize only slightly in, water. Ammonia, NH_3, is the most common example. See Appendix G.

5   Solubility Guidelines for Compounds in Aqueous Solutions

The strong electrolytes are (1) the strong acids, (2) the strong soluble bases, and (3) *most soluble ionic salts*. To determine whether a salt is water-soluble or not, you should learn the **solubility guidelines in Section 6-2, part 5** in the text.

1.   The common inorganic acids are soluble in water. Low-molecular-weight organic acids are also soluble.

2.   All common compounds of the Group IA metal ions (Li^+, Na^+, K^+, Rb^+, Cs^+) and the ammonium ion, NH_4^+, are soluble in water.

3.   The common nitrates, NO_3^-; acetates, CH_3COO^-; chlorates, ClO_3^-; and perchlorates, ClO_4^-, are soluble in water.

4.   (a) The common chlorides, Cl^-, are soluble in water except $AgCl$, Hg_2Cl_2, and $PbCl_2$.
      (b) The common bromides, Br^-, and iodides, I^-, show approximately the same solubility behavior as chlorides, but there are some exceptions. As these halide ions (Cl^-, Br^-, I^-) increase in size, the solubilities of their slightly soluble compounds decrease.
      (c) The common fluorides, F^-, are soluble in water except MgF_2, CaF_2, SrF_2, BaF_2 and PbF_2.

5.   The common sulfates, SO_4^{2-}, are soluble in water except $PbSO_4$, $BaSO_4$, and $HgSO_4$; $CaSO_4$, $SrSO_4$, and Ag_2SO_4 are moderately soluble.

6.   The common metal hydroxides, OH^-, are *insoluble* in water except those of the Group IA metals and the heavier members of the Group 2A metals, beginning with $Ca(OH)_2$.

7.   The common carbonates, CO_3^{2-}, phosphates, PO_4^{3-}, and arsenates, AsO_4^{3-}, are insoluble in water except those of the Group IA metals and NH_4^+. $MgCO_3$ is moderately soluble.

8.   The common sulfides, S^{2-}, are insoluble in water except those of the Group IA and Group 2A metals and the ammonium ion.

Examples of soluble salts and their *dissociations* in aqueous solutions are

Most soluble salts are completely, or nearly completely, *dissociated* in dilute aqueous solutions. There are some exceptions.

6-2   Reactions in Aqueous Solutions (Study Tables 6-4 and 6-5)

1.   **Formula unit equations** show complete formulas for all compounds. When metallic zinc is dropped into a solution of copper(II) sulfate it dissolves as metallic copper precipitates.

$$Zn\;(s)\;+\;CuSO_4\;(aq)\;\rightarrow\;ZnSO_4\;(aq)\;+\;Cu\;(s)$$

2.   In **total ionic equations,** formulas are written to show the predominant form in which each substance exists when it is in contact with, or dissolved in, aqueous solution. Both $CuSO_4$ and $ZnSO_4$ are soluble ionic compounds (*not* molecules).

$$Zn(s)\;+\;Cu^{2+}\,(aq)\;+\;SO_4{}^{2-}\,(aq)\;\rightarrow\;Zn^{2+}\,(aq)\;+\;SO_4{}^{2-}\,(aq)\;+\;Cu(s)$$

3.   A **net ionic equation** shows only the species that react. The sulfate ions are called "spectator" ions because they do not participate in the reaction. We cancel them to get the net ionic equation

$$Zn(s)\;+\;Cu^{2+}\,(aq)\;\rightarrow\;Zn^{2+}\,(aq)\;+\;Cu(s)$$

To determine whether the formula for a substance should be written in ionic form in ionic equations, you must answer two questions.

1.   Is it soluble in water?

2.   If it is soluble, is it mostly ionized or dissociated in water?

If *both* answers are yes, it is written in ionic form.

If *either* answer is no, it is *not* written in ionic form.

The common substances that are written in ionic form in ionic equations are (1) strong acids, (2) strong soluble bases, and (3) soluble ionic salts.

NAMING SOME INORGANIC COMPOUNDS

6-3 Naming Some Binary Compounds

Binary compounds contain *two* elements. In binary compounds, the more metallic element is named first. The less metallic element is named last with an **-ide** ending. An unambiguous **"stem"** is taken from the name of the more metallic element so that the suffix **-ide** can be attached to the stem.

B – bor	C – carb	N – nitr	O – ox	H – hydr
	Si – silic	P – phosph	S – sulf	F – fluor
		As – arsen	Se – selen	Cl – chlor
		Sb – antimon	Te – tellur	Br – brom
				I – iod

Consider binary ionic compounds that contain a **metal cation** *that exhibits only one oxidation state* (other than zero) *and a* **nonmetal anion**.

Compound	Name	Compound	Name
LiBr	_____	Al_2O_3	_____
$MgCl_2$	_____	Na_3P	_____
Li_2S	_____	Mg_3N_2	_____

Binary ionic compounds containing *metals that exhibit more than one oxidation state* combined with *nonmetals* are named by two systems. In the older system **-ic,** and **-ous** are used as suffixes to distinguish between higher and lower oxidation states, respectively. Two suffixes make distinctions between only two oxidation states. In the modern system (Stock system), Roman numerals are used to indicate oxidation states for *metals that exhibit variable oxidation states.*

Compound	Older Name	Modern Name
$FeBr_2$	_____	_____
$FeBr_3$	_____	_____
SnO	_____	_____
SnO_2	_____	_____
$TiCl_2$	_____	_____

88

Compound	Older Name	Modern Name
$TiCl_3$	_____	_____
$TiCl_4$	_____	_____

Pseudobinary ionic compounds consist of polyatomic ions that behave as simple binary ionic compounds. There are three common polyatomic ions.

OH^- hydroxide CN^- cyanide NH_4^+ ammonium

These compounds are named using the binary ionic naming system but including these polyatomic ion names.

KOH   _____

$Ba(OH)_2$ _____

$Al(OH)_3$ _____

$Fe(OH)_2$ _____

$Fe(OH)_3$ _____

$Ba(CN)_2$ _____

$(NH_4)_2S$ _____

NH_4CN _____

Binary molecular compounds are mostly two nonmetals bonded together. The oxidation numbers are not indicated by Roman numbers or suffixes. The proportions of the binary molecular compounds are indicated by a prefix system that is used for both elements. These prefixes are *mono-*, *di-*, *tri-*, *tetra-*, *penta-*, *hexa-*, *hepta-*, *octa-*, *nona-*, and *deca-*. The prefix "mono-" is only used in CO (carbon monoxide). The final "a" of a prefix is not used when the nonmetal stem begins with the letter "o."

These compounds can be named using the binary molecular compound naming system.

SO_2 _____

CS_2 _____

SO_3 _____

SF_4 _____

Binary acids are compounds in which hydrogen and another nonmetal (other than hydrocarbon compounds) are named **hydrogen (stem)ide.** Aqueous solutions of many binary compounds that contain hydrogen and another nonmetal are acids. They are named **hydro(stem)ic** acid. Refer to Appendix F for more examples.

Compound	Name	Aqueous Solution (aq)
HF	_____	_____
HCl	_____	_____
H_2S	_____	_____
H_2Se	_____	_____

6-4   Naming Ternary Acids and Their Salts

Ternary compounds contain three elements. Ternary acids (oxoacids, sometimes called oxyacids) contain hydrogen, oxygen, and (in most cases) another nonmetal. A ternary acid for each central nonmetal is called the **-ic** acid. Study the list of "ic" acids given below.

3A	4A	5A	6A	7A
H_3BO_3 boric acid	H_2CO_3 carbonic acid	HNO_3 nitric acid		
	H_4SiO_4 silicic acid	H_3PO_4 phosphoric acid	H_2SO_4 sulfuric acid	$HClO_3$ chloric acid
		H_3AsO_4 arsenic acid	H_2SeO_4 selenic acid	$HBrO_3$ bromic acid
			H_6TeO_6 telluric acid	HIO_3 iodic acid

Two ternary acids containing the same central nonmetal are named **(stem)ic acid** to indicate the higher oxidation state, and **(stem)ous acid** to indicate the lower oxidation state of the central nonmetal. Anions of **-ous** acids have **-ite** suffixes. Anions of **-ic** ternary acids have **-ate** suffixes.

Acid	Name	Sodium Salt	Name
HNO_2	_____	$NaNO_2$	_____
HNO_3	_____	$NaNO_3$	_____
H_2SO_3	_____	Na_2SO_3	_____
H_2SO_4	_____	Na_2SO_4	_____
$HClO_2$	_____	$NaClO_2$	_____
$HClO_3$	_____	$NaClO_3$	_____

When there are more than two ternary acids of a central nonmetal, two *prefixes*, **hypo-** to indicate the lowest oxidation state, and **per-** to indicate the highest oxidation state, are used with the **-ous** and **-ic** suffixes. The prefixes are retained in the salt's name.

Acid	Name	Sodium Salt	Name
$HClO$	_____	$NaClO$	_____
$HClO_2$	_____	$NaClO_2$	_____
$HClO_3$	_____	$NaClO_3$	_____
$HClO_4$	_____	$NaClO_4$	_____

Ternary salts are compounds that result from replacing the hydrogen in a ternary acid with another ion. Salts contain the cation of a base and the anion of an acid. The common cations and anions are listed in Table 6-6. The cation is named first and the anion is named last.

Acidic salts contain one or more acidic (ionizable) hydrogen atoms. In the older system of nomenclature, the prefix bi- indicated the presence of an acidic hydrogen atom in a salt. In the modern system, prefixes are used to indicate the number of acidic hydrogen atoms.

91

Acidic Salt	Old System	Modern System
$NaHCO_3$		
$KHSO_4$		
KH_2PO_4		
K_2HPO_4		

CLASSIFYING CHEMICAL REACTIONS

Because millions of chemical reactions are known, we find it useful to classify reactions into types. We classify them as (1) oxidation-reduction reactions, (2) combination reactions, (3) decomposition reactions, (4) displacement reactions, and (5) metathesis reactions. As we shall see, many reactions fit into more than one class, and metathesis reactions can be divided into several subclasses of reactions.

6-5 Oxidation-Reduction Reactions: Introduction

Oxidation and reduction always occur simultaneously. Redox reactions are "energy producing reactions."

Oxidation	-	an algebraic increase in oxidation number
	-	the process in which electrons are (or appear to be) lost
Reduction	-	an algebraic decrease in oxidation number
	-	the process in which electrons are (or appear to be) gained
Oxidizing agents	-	substances that gain electrons and oxidize other substances
	-	oxidizing agents are always reduced.
Reducing agents	-	substances that lose electrons and reduce other substances
	-	reducing agents are always oxidized.

Example 6-1:   Write and balance the formula unit, total ionic, and net ionic
              equations for the oxidation of sulfurous acid to sulfuric acid by
              oxygen in acidic aqueous solution.

formula unit

total ionic

net ionic

A disproportionation reaction is an oxidation-reduction reaction in which the same element is oxidized and reduced.

6-6   Combination Reactions

Combination reactions involve the combination of two substances to form a compound. They can consist of an element combined with an element to form a compound, a compound and an element to form a compound, or two compounds to form a compound.

1   Element + Element → Compound

Metal + Nonmetal → Binary Ionic Compound

Most *metals* react with most *nonmetals* to form ionic binary compounds. The following are examples. There are more examples of metals reacting with nonmetals in section 7-2.

Nonmetal + Nonmetal → Covalent Binary Compound

When *nonmetals* react with each other they form binary *covalent* compounds. A higher oxidation state of a nonmetal is formed when it reacts with an excess of another nonmetal. Consider the reactions of arsenic (Group VA) with a limited amount of chlorine *and* with an excess of chlorine.

Note: The *common* oxidation numbers of nonmetals in odd numbered periodic groups are odd, while those in even numbered periodic groups are even.

Consider the reaction of selenium (Group VIA) with limited and with excess fluorine.

2   Compound + Element → Compound

There are many examples of common reactions in this category. Arsenic trichloride reacts with chlorine to produce arsenic pentachloride.

Sulfur tetrafluoride reacts with fluorine to form sulfur hexafluoride.

Among the most common of these reactions are *reactions of oxygen with oxides of nonmetals* to form oxides in which the other nonmetal is in a higher oxidation state. Carbon monoxide combines with oxygen to form carbon dioxide.

Sulfur dioxide combines with oxygen to form sulfur trioxide. A catalyst and high temperatures are required.

Tetraphosphorus hexoxide combines with oxygen to form tetraphosphorus decoxide.

94

3   Compound + Compound → Compound

Examples include: The reaction of gaseous ammonia with gaseous hydrogen chloride to form solid ammonium chloride, a salt.

The reaction of lithium oxide with sulfur trioxide to form lithium sulfate, a salt.

6-7   Decomposition Reactions

We can classify decomposition reactions into three categories.

1   Compound → Element + Element

The thermal decomposition of gaseous dinitrogen oxide gives

The electrolytic decomposition of molten calcium chloride gives

The photochemical decomposition of silver halides in the presence of light to form small granules of dark metallic silver and free halogens. This reaction is responsible for the darkening of photographic negatives upon exposure.

2   Compound → Compound + Element

The decomposition of hydrogen peroxide is catalyzed by light as well as by many metalions.

3   Compound → Compound + Compound

The thermal decomposition of ammonium hydrogen carbonate gives

6-8   Displacement Reactions

These are reactions in which one element displaces another element from a compound. The displacement of copper from copper(II) sulfate by zinc is an example. Active metals displace less active metals or hydrogen from their compounds (Table 6-9).

1   [More Active Metal   [Less Active Metal +
     + Salt of Less Active Metal]   →   Salt of More Active Metal]

Metallic copper will displace silver from silver nitrate in aqueous solution.

formula unit

total ionic

net ionic

2   [Active Metal + Nonoxidizing Acid] → [Hydrogen + Salt of Acid]

Metallic aluminum reacts with sulfuric acid solution to produce gaseous hydrogen and a solution of soluble, ionic aluminum sulfate, i.e., aluminum displaces hydrogen.

formula unit

total ionic

net ionic

96

This net ionic equation represents the reaction that occurs whenever metallic Al is dropped into a solution of *any strong, nonoxidizing acid*. (Nitric acid, HNO_3, is the common strong *oxidizing* acid; its reaction produces oxides of nitrogen such as NO or NO_2 rather than H_2.) Some common metals that are active enough to displace H^+ from acids include K, Ca, Na, Mg, Al, Zn, Fe, Sn, and Pb (Table 6-9).

3 **[Active Nonmetal +**   **[Less Active Nonmetal +**
  Salt of Less Active Nonmetal] $\rightarrow$ **Salt of More Active Nonmetal]**

Many nonmetals displace less active nonmetals from combination with a metal or other cation. Each halogen will displace less active (heavier) halogens from their binary salts. Consider the reaction that occurs when chlorine is bubbled into a solution of sodium iodide.

formula unit equation

total ionic equation

net ionic equation

6-9 Metathesis Reactions

Metathesis reactions are reactions in which the ions of two compounds exchange partners. *No changes in oxidation number occur.* They are sometimes called "double displacement" reactions.

$$AX + BY \rightarrow AY + BX$$

1 **Acid-Base (Neutralization) Reactions: Formation of a Nonelectrolyte**

The reaction of an acid with a metal hydroxide base produces a salt and water. Such reactions are called **neutralization reactions** because the typical properties of acids and bases are neutralized. In a *neutralization reaction* H^+ ions and OH^- ions combine to form water molecules. Consider the reaction of hydrobromic acid with potassium hydroxide.

formula unit

total ionic

net ionic

Consider the reaction of calcium hydroxide with nitric acid.

formula unit

total ionic

net ionic

The generalized reaction of a weak *monoprotic* acid with a strong soluble base can be represented as

2   Precipitation Reactions

In *precipitation reactions* an insoluble compound or *precipitate* is formed as a product. An example is the precipitation of calcium carbonate by mixing solutions of the soluble ionic salts, calcium nitrate and potassium carbonate.

formula unit

total ionic

net ionic

Consider the reaction of calcium chloride with sodium phosphate.

formula unit

total ionic

net ionic

6-10   Gas-Formation Reactions

A gas-formation reaction is a type of reaction in which there is a formation of an insoluble or slightly soluble gas when there are no gaseous reactants. Displacement reactions in which an active metal displaces from an acid or from water are gas-formation reactons; they are not methathesis reactions. Consider hydrochloric acid with calcium carbonate to form carbonic acid.

formula unit

total ionic

net ionic

6-11   Summary of Reaction Types

We have classified many chemical reactions. Many reactions can fit into more than one category.

As you study Table 6-11 carefully, you will develop a better understanding of why we classify reactions into the various categories. Imagine how difficult it would be to try to learn millions of individual reactions!

Synthesis Question

Barium sulfate is a commonly used imaging agent for gastrointestinal X-rays. This compound can be prepared by some of the kinds of reactions described in this chapter. Write a balanced equation for a reaction in aqueous solution for the production of barium sulfate. You may choose any aqueous starting materials that will form barium sulfate!

Group Activity

Pretend that you are a lab TA and that you have been given the assignment to prepare unknowns for a qualitative analysis experiment. In a single solution you must have the following cations: Bi^{3+}, Cd^{2+}, and Cu^{2+}. You must make this solution using soluble salts that contain three different anions. Which three compounds would you choose to make this solution so that no precipitate forms?

Chapter Seven

CHEMICAL BONDING

The attractive forces that hold atoms together in compounds are called chemical bonds. Chemical bonds are formed when two or more atoms combine to form a compound. When a compound decomposes into its constituent elements, chemical bonds are broken. The electrons involved in bonding are usually those in the outermost (valence) shell.

(1) *Ionic bonding* results from electrostatic interactions among ions, which are formed by the net transfer of one or more electrons from one atom or group of atoms to another.

(2) *Covalent bonding* results from sharing one or more electron pairs between two atoms.

Let us compare some properties of ionic and covalent compounds. There is no sharp dividing line between ionic and covalent bonding (compounds). The descriptions given below are generally applicable.

	Property	Ionic Compounds	Covalent Compounds
1.	Melting Point	_____	_____
2.	Solubility in polar solvents such as water	_____	_____
3.	Solubility in nonpolar solvents such as octane	_____	_____
4.	Electrical conductivity of molten (liquid) compounds	_____	_____
5.	Electrical conductivity of aqueous solutions	_____	_____

7-1 Lewis Dot Formulas of Atoms

The Lewis dot representation (or Lewis dot formulas) provide a convenient bookkeeping method for *valence electrons*, the electrons that are involved in chemical bonding. See Table 7-1.

Examples are

H   He

Li   Be   B   C   N   O   F   Ne

Na   Mg   Al   Si   P   S   Cl   Ar

IONIC BONDING

7-2 Formation of Ionic Compounds

An ion is an atom, or group of atoms, that carries an electrical charge. Ionic bonding is the attraction of oppositely charged ions in large numbers to form a solid. Atoms that *lose* electrons become *positively charged* and are called *cations*. Atoms that *gain* electrons become *negatively charged* and are called *anions*.

Group IA Metals and Group 7A Nonmetals

Ionic bonding occurs when elements with low ionization energies (*metals*) react with elements with high electronegativities (*nonmetals*). All of the IA metals react with all of the Group 7A nonmetals to form ionic compounds. Lithium reacts with fluorine to form lithium fluoride, LiF, an ionic compound. The equation for the reaction is

The formula for lithium fluoride is LiF because each lithium atom loses one electron while each fluoride atom gains one electron. We may represent the reaction in more detail as

We can also use Lewis formulas to represent the neutral atoms and the ions they form.

Lithium atoms lose one electron each to form lithium ions, Li$^+$. Lithium ions contain two electrons, the same number as helium, the preceding noble gas. Lithium ions are *isoelectronic* with helium. Fluorine atoms gain one electron each to form fluoride ions, F$^-$, which contains ten electrons. This is the same number as neon, the following noble gas. Fluoride ions are *isoelectronic* with neon. *Isoelectronic* species contain the same number of electrons.

Potassium reacts with bromine to form potassium bromide, KBr, an ionic compound. The equation for the reaction is

In more detail we represent the reaction as

The appropriate Lewis dot formulas are

Potassium ions are isoelectronic with argon, the preceding noble gas. Bromide ions are isoelectronic with krypton, the following gas.

We may represent the reactions of the Group IA metals with the Group 7A nonmetals in general terms (to describe 20 reactions) as

$$M = Li \rightarrow Cs$$
$$X = F \rightarrow I$$

In more detail the generalized reaction may be represented as

Representative elements that form binary *ionic compounds* usually form ions that have noble gas configurations. *Metals lose electrons* to attain the configuration of the *preceding* noble gas. *Nonmetals gain electrons* to attain the configuration of the *following* noble gas.

Ionic compounds consist of extended arrays of oppositely charged ions. They have relatively high melting points because of the strong electrostatic attractions among oppositely charged ions. Coulomb's Law is

q^- = magnitude of charge on anions

q^+ = magnitude of charge on cations

F = force of attraction between oppositely charged ions

Smaller ions with (opposite) higher charges attract each other more strongly than larger ions with (opposite) lower charges.

We refer to *formula units* of ionic compounds because there are no molecules of these substances. See Figure 7-1, the structure of sodium chloride, NaCl.

Group IA Metals and Group 6A Nonmetals

Lithium, a Group IA metal, reacts with oxygen, a Group 6A nonmetal, to form lithium oxide, Li_2O, an ionic compound. The equation for this reaction is

The formula for lithium oxide is Li_2O because each lithium atom loses one electron while each oxygen atom gains two. In more detail, this reaction may be represented as

The Lewis dot formulas are

Example 7-1:   Arrange the following compounds in order of increasing forces of attractions among ions. KCl, Al_2O_3, CaO

The melting points of these compounds increase in the same direction.

Group 2A Metals and Group 6A Nonmetals

Magnesium, a Group 2A metal, reacts with oxygen, a Group 6A nonmetal, to form magnesium oxide, MgO, an ionic compound. The equation for this reaction is

MgO is the formula for magnesium oxide because the magnesium atom loses two electrons while the oxygen atoms gains two electrons. The reaction may also be represented as

The Lewis dot formulas are

d-Transition Metal Ions

The d-transition metals in their atomic electron configurations have both s and d electrons. When these metals form ions the s electrons are lost s before the d electrons. The two common ions of iron, Fe^{2+} and Fe^{3+}, are examples

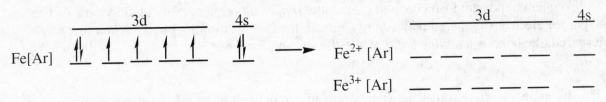

Binary Ionic Compounds: A Summary

We may summarize the reactions of the representative metals with the representative nonmetals to form binary ionic compounds. In the following table, M represents *metals* and X represents *nonmetals* in the indicated group in the periodic table. Study Table 7-2 carefully for more details.

Groups			General Formula	Specific Example
IA	+	VIIA	MX	
IIA	+	VIIA	MX_2	
IIIA	+	VIIA	MX_3	
IA	+	VIA	M_2X	
IIA	+	VIA	MX	
IIIA	+	VIA	M_2X_3	
IA	+	VA	M_3X	
IIA	+	VA	M_3X_2	
IIIA	+	VA	MX	

COVALENT BONDING

All ionic compounds are solids that have high melting points, compared to covalent compounds. In contrast, covalent compounds are gases, liquids, or low melting solids. A few general properties of ionic and covalent compounds were contrasted on page 251.

A covalent bond is formed when two atoms share one or more pairs of electrons. Covalent bonding occurs when the electronegativity difference between elements (atoms) is zero or relatively small.

7-3   Formation of Covalent Bonds

Covalent bonds are formed when two nonmetals react with each other. Some binary compounds that contain a metal and a nonmetal are also covalent.

Covalent bonds are formed when two atoms *share* one or more pairs of electrons. When *one* pair of electrons is shared between two atoms, the bond is a *single* bond. When *two pairs* are shared, the bond is a *double bond*, while the sharing for *three electron pairs* results in a *triple bond*.

The main attractive forces between atoms in covalent compounds are electrostatic in nature. The variation in the potential energy with internuclear distance for diatomic molecules such as hydrogen, H_2, is

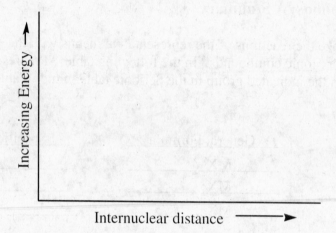

For every diatomic molecule there is an internuclear separation that corresponds to the minimum in a potential energy diagram. This distance is the internuclear separation in the molecule, i.e., the distance between centers of nuclei. See Figures 7-3 and 7-4.

We can represent the formation of an H_2 molecule from two H atoms by Lewis dot formulas.

The formation of an HCl molecule from H and Cl atoms is represented as

A *single covalent bond* is formed when two atoms share one pair of electrons, a *double covalent bond* is formed when two atoms share two electron pairs, and a *triple covalent bond* is formed when two atoms share three electron pairs.

7-4   Bond Lengths and Bond Energies

The *bond length* of a covalent bond is the distance in which the attractive and repulsive forces are balanced. The combination of bonded atoms at this distance is more stable (lower energy) than if the atoms were separated. The difference in this energy is known as the *bond dissociation energy* (bond energy). See Table 7-3 and Table 7-4.

7-5   Lewis Formulas for Molecules and Polyatomic Ions

A Lewis dot formula shows the electrons in the outermost valence shells of all the atoms in a molecule or ion. The Lewis dot and dash formulas for some *homonuclear* diatomic molecules are

The Lewis dot and dash formulas for the hydrogen halides (*heteronuclear* diatomic molecules) are

The Lewis dot and dash formulas for water, ammonia, and the ammonium ion are

7-6   Writing Lewis Formulas: The Octet Rule

In the molecules for which we have drawn Lewis dot formulas, all the atoms of representative elements achieved noble gas configurations by sharing electrons. All these atoms except H attained a share of eight electrons in their outer shells (the noble gas configuration). H attained a share of two electrons (the He configuration). The *octet rule* is: **Representative elements achieve noble gas configurations in most of their compounds.**

A Guide to Writing Lewis Dot Formulas

The following relationship is a simple aid in writing dot formulas for molecules and polyatomic ions that contain *only representative elements*, S = N - A.

S = Total number of e- **SHARED**

N = Total number of e- **NEEDED** to attain noble gas configurations

A = Total number of e- **AVAILABLE** in the outer shells of all atoms (sum of periodic group numbers of all atoms, using 8 rather than 0 for noble gases).
   a) One e- must be added for each negative charge and
   b) One e- must be subtracted for each positive charge on ions.

1) The central atom in a molecule or polyatomic ion is the one that requires the largest number of electrons to achieve a noble gas configuration.

2) Hydrogen is never the central atom.

3) When one must decide which of two elements from the same group in the periodic table is the central atom, the less electronegative atom is usually chosen.

The rules are outlined and illustrated beginning on page 264 in your text.

Example 7-2:   Write Lewis dot and dash formulas for hydrogen cyanide, HCN.

Example 7-3:   Write Lewis dot and dash formulas for the sulfite ion, SO_3^{2-}.

Example 7-4:   Write Lewis dot and dash formulas for sulfur trioxide, SO_3.

7-7 Formal Charges

Formal charge is a commonly used method to help determine the correct Lewis formulas of some species. Formal charge is the **hypothetical** charge on an atom in a *molecule or polyatomic ion*. The Lewis formula that gives each atom a formal charge of zero or very nearly zero is the best choice.

Rules for Assigning Formal Charges

Formal charge on an atom in a Lewis formula is

1.   Formal charge = (group number) - [(number of bonds) + (number of unshared e)]

2.   In Lewis formulas, an atom that has the same number of bonds as its periodic group number has a formal charge of 0.

3.   a.   In molecules, the formal charges must sum to 0.

      b.   In polyatomic ions, the formal charges must sum to the ions' charage.

Nitrosyl chloride, CINO, is a good example of a compound that two acceptable Lewis formulas can be distinguished using formal charge.

7-8   Writing Lewis Formulas: Limitations of the Octet Rule
(Species to which S = N - A cannot be applied)

Species that violate the octet rule, and therefore species for which S = N - A must be modified, include

1.   Most covalent compounds of Be.

2.   Most covalent compounds of Group IIIA elements.

3.   Compounds or ions containing an odd number of electrons.

4.   Species in which the central element must have a share in more than 8 valence electrons to accommodate all substituents.

In cases where the octet rule does *not* apply, the substituents attached to the central atom nearly always attain noble gas configurations. The central atom does not.

Example 7-5:   Write dot and dash formulas for boron tribromide, BBr_3.

Example 7-6:   Write dot and dash formulas for arsenic pentafluoride, AsF$_5$.

7-9   Resonance

The dot (dash) formula shown above for SO$_3$ (example 7-4) is not the only acceptable one. The double bond could equally well involve either of the other two oxygen atoms.

When two or more dot representations are necessary to describe the bonding in a molecule or polyatomic ion, the molecule or polyatomic ion is said to exhibit *resonance*. Double-headed arrows ($\leftrightarrow$) are used to indicate resonance formulas. The actual bonding in SO$_3$ is intermediate among the forms shown in the three resonance formulas above. The S-O bond distance is the same for all three S-O bonds in SO$_3$ molecules. It is intermediate in length and bond strength between a single and a double sulfur-oxygen bond. The SO$_3$ molecule is more accurately represented as

Drawing the SO$_3$ molecule this way shows delocalization.

7-10 Polar and Nonpolar Covalent Bonds

Covalent bonds are classified as either polar or nonpolar. In *nonpolar covalent bonds* the electron pair(s) is (are) shared equally between two nuclei. For an electron pair to be shared equally between two atoms, the two atoms must have the same electronegativity. True nonpolar covalent bonds exist only in homonuclear molecules such as H$_2$, O$_2$, N$_2$, F$_2$, and Cl$_2$.

112

Nonpolar bonds are covalent bonds in which there is a *symmetrical distribution* of electron density between two nuclei. In *polar covalent bonds*, an electron pair is shared unequally between two nuclei. Polar covalent bonds are formed when two atoms with different electronegativities share one or more electron pairs. Most covalent bonds are polar. Consider the hydrogen halides, HF, HCl, HBr, and HI.

7-11 Dipole Moments

The dipole moment of a molecule, μ, is the product of distance, d, separating charges of equal magnitude and opposite sign and the magnitude of the charge, q. Nonpolar molecules have zero dipole moments. Molecules with small separations of charge have small dipole moments. Molecules with large separations of charge (very unsymmetrical electron distributions) have large dipole moments. See Table 7-3.

Dipole moments for *entire molecules* can be determined experimentally. See Figure 7-5. Many nonpolar molecules have polar bonds. For a molecule to be polar *both* of the following conditions must be met.

1. There must be at least one polar bond present or at least one lone pair of electrons on the central atom.

2. The polar bonds, if there are more than one, and lone pairs must be arranged so that their dipole moments do *not* cancel one another.

7-12 The Continuous Range of Bonding Types

Polar covalent bonds may be considered as intermediate between pure (nonpolar) covalent bonds and pure ionic bonds. This bond polarity is sometimes described in terms of *partial ionic character*. Calculations based upon this approach and the measured dipole moment describes the bond as a mixture of pure covalent character and pure ionic character. Such calculations for gaseous HCl describe the HCl bond as having about 17% "ionic character".

When cations and anions interact strongly, some amount of electron sharing takes place; in such cases we can consider the ionic compound as having some *partial covalent character*. When this approach is used, nearly all bonds can be described as having both some partial covalent character and some partial ionic character.

The polarity of a bond is related to the electronegativity difference between two bonded atoms.

A molecule in which the centers of positive and negative charge do not coincide is said to possess a *dipole*.

Synthesis Question

As we all know, in the winter time we are more likely to be shocked when we walk across carpet and touch a door knob. Here is another wintertime experiment to perform. Turn on a water faucet until you have a continuous but small stream of water coming from the faucet. Brush your hair vigorously, and then hold the brush near the stream of water. You will notice that the stream bends towards the brush. Why does the stream of water bend?

Group Question

On a recent "infomercial" it was claimed that placing a small horseshoe magnet over the fuel intake line to your car's carburetor would increase fuel mileage by 50%. The reason given for the mileage increase was that "the magnet aligned the molecules causing them to burn more efficiently." Will this work? Should you buy this product?

Chapter Eight

MOLECULAR STRUCTURE AND COVALENT BONDING THEORIES

The structures of many compounds have been determined experimentally. Our theories and predictions must be consistent with *experimentally determined* facts, i.e., the *known* structures of compounds. We shall discuss two theories that provide insight into covalent bonding. The two theories go hand-in-hand.

8-1   A Preview of the Chapter

In this chapter we shall apply two theories of chemical bonding and structure. These two theories are the *valence shell electron pair repulsion (VSEPR) theory* which allows us to predict molecular shapes and the *valence bond (VB) theory* which lets us describe how the atomic orbitals are used in making bonds. The following provides a procedure for the analysis of the structure and bonding in any compound.

1.   Draw the Lewis electron dot formula for the molecule or polyatomic ion, and identify a *central atom*—an atom that is bonded to more than one other atom. (Section 8-2)

2.   Count the number of regions of high electron density on the central atom. (Section 8-2)

3.   Apply the VSEPR theory to determine the arrangement of the regions of high electron density (the electronic geometry) about the central atom. (Tables 8-1 & 8-4)

4.   Using the Lewis formula as a guide, determine the arrangement of the *bonded atoms* (the *molecular geometry*) about the central atom, as well as the location of the unshared valence electron pairs on that atom (parts B of Sections 8-5 through 8-12; Tables 8-3 and 8-4). This description includes ideal bond angles.

5. If there are lone pairs of electrons on the central atom, consider how their presence might modify somewhat the *ideal* molecular geometry and bond angles deduced in step 4 (Section 8-2; Sections 8-8 through 8-12)

6. Use the VB theory to determine the kinds of *hybrid orbitals* utilized by the central atom; describe the overlap of these orbitals to form bonds and the orbitals that contain unshared valence shell electron pairs on the central atom (parts C of Sections 8-5 through 8-12; Sections 8-13; 8-14; Tables 8-2 and 8-4).

7. If more than one atom can be identified as a central atom, repeat steps 2 through 6 for each central atom, to build up a picture of the geometry and bonding in the entire molecule or ion.

8. When all central atoms have been accounted for, use the entire molecular geometry, electronegativity differences, and the presence of lone pairs of valence shell electrons on the central atom to predict *molecular polarity* (Section 8-3; parts B of Sections 8-5 through 8-12).

8-2   Valence Shell Electron Pair Repulsion Theory

The basic idea of VSEPR *theory* is that regions of high electron density around the central atom are as far apart as possible to minimize repulsions among them (electron pairs). The geometry of a molecule or polyatomic ion is determined by the locations of regions of high electron density around the central atom(s), the *electronic geometry*. In determining electronic geometry a single bond, a double bond, a triple bond, and an unshared pair of electrons (lone pair) are each considered as a single region of high electron density. Most simple molecules and polyatomic ions have two, three, four, five, or six regions of high electron density around the central atom. See Table 8-1.

# of regions of high electron density	Geometric Name	Bond Angles	Line Drawing

118

The *molecular geometry* (*ionic geometry* for polyatomic ions) refers to the locations of atoms (not just electron pairs) around the central atom in a molecule or polyatomic ion. This contrasts to the *electronic geometry* which refers to the locations of the electron pairs around the central atom in a molecule or polyatomic ion.

Methane, CH_4, has tetrahedral electronic geometry *and* tetrahedral molecular geometry because an H atom is associated with each of the four electron pairs around the C atom.

Water has four pairs of electrons around the O atom, but only two H atoms. H_2O has tetrahedral electronic geometry, but *bent* or *angular* molecular geometry.

Unshared pairs (lone pairs) of electrons (lp) require a larger volume than bonding pairs (bp) of electrons. The relative magnitudes of repulsive forces among pairs of electrons around the central atom are

8-3   Polar Molecules: The Influence of Molecular Geometry

For a molecule to be polar, *both* of the following conditions must be met.

1.   There must be at least one polar bond *or* one lone pair of electrons on the central atom.

and

2.   a.   The polar bonds, if there are more than one, must *not* be so symmetrically arranged that their bond polarities cancel.

or

b.   If there are more than one lone pair of electrons on the central atom, they must *not* be so symmetrically arranged that their polarities cancel.

8-4   Valence Bond (VB) Theory

We shall explain covalent bonding in terms of Valence Bond Theory to account for *observed* (experimentally determined) structures of molecules and polyatomic ions. It postulates that covalent bonds are formed when orbitals on two different atoms overlap so that electrons are shared in the regions of orbital overlap. Hybridization, or mixing of atomic orbitals on the central atom, is often postulated to account for the shapes of molecules and polyatomic ions.

119

Regions of High Electron Density	Electronic Geometry	Hybridization

We now describe the structures and bonding that characterize several important classes of molecules and polyatomic ions.

In the next sections we will use the following terminology

A = central atom

B = bonding pair of electrons around central atom

U = lone pair of electrons around central atom

For example: AB_3U designates that there are 3 bonding pairs and 1 lone pair of electrons around the central atom.

MOLECULAR SHAPES AND BONDING

8-5 Linear Electronic Geometry: AB_2 Species (No Lone Pairs of Electrons on A)

Examples include BeI_2, $BeBr_2$, $BeCl_2$, $HgCl_2$, and $CdCl_2$. All are *linear, nonpolar* molecules. Consider $BeCl_2$.

120

Electronic Structures of Atoms

Dot Formulas   Electronic Geometry about Be

VSEPR   Polarity

Molecular Geometry

Valence Bond Theory (Hybridization)

8-6 Trigonal Planar Electronic Geometry: AB3 Species (No Lone Pairs of Electrons on A)

Examples include BF_3, BCl_3, etc. All are *trigonal planar, nonpolar* molecules. Consider BCl_3.

Electronic Structures of Atoms

Dot Formulas   Electronic Geometry about B

VSEPR   Polarity

Molecular Geometry

Valence Bond Theory (Hybridization)

8-7   Tetrahedral Electronic Geometry: AB₄ Species
         (No Lone Pairs of Electrons on A)

Examples include CH_4, CF_4, CCl_4, SiH_4, SiF_4, etc. All are *tetrahedral, nonpolar* molecules. Consider SiH_4.

Electronic Structures of Atoms

Dot Formulas   Electronic Geometry about C

VSEPR   Polarity

Molecular Geometry

Valence Bond Theory (Hybridization)

The series of saturated hydrocarbons, known as alkanes, have the general formula C_nH_{2n+2}. They are CH_4, C_2H_6 (H_3C-CH_3), C_3H_8 (H_3C-CH_2-CH_3), etc. Each C atom is located at the center of a tetrahedron. Beyond methane, CH_4, each alkane can be thought of as a chain of interlocking tetrahedra with a C atom at the center of each tetrahedron, and enough H atoms to form a total of four bonds for each C atom.

Methane

Ethane

Propane

These saturated hydrocarbons are discussed in detail in Chapter 27.

8-8   Tetrahedral Electronic Geometry: AB₃U Species
(One Lone Pair of Electrons on A)

Two examples are NH_3, and NF_3. Both are *pyramidal, polar molecules.* These are our first examples of molecules that have lone pairs of electrons on their central atoms. These molecules have different electronic and molecular geometries. Furthermore, when all 3 substituents are the same, the molecules are *polar*, unlike AB₃ molecules.

123

Electronic Structures of Atoms

Dot Formulas

VSEPR

Molecular Geometry about N

Polarities

NH$_3$

NF$_3$

Valence Bond Theory (Hybridization)

We see that AB$_3$U molecules with one unshared electron pair on A (e.g., NH$_3$ and NF$_3$) are quite different from AB$_3$ molecules with *no* unshared electron pairs on A (e.g., BF$_3$ and BCl$_3$).

8-9   Tetrahedral Electronic Geometry: AB_2U_2 Species
(Two Lone Pairs of Electrons on A)

Water (H_2O) is the most important molecule of this type. It is an *angular, polar molecule*.

Electronic Structures of Atoms

Dot Formulas   Electronic Geometry about O

VSEPR   Polarity

Molecular Geometry

Valence Bond Theory (Hybridization)

8-10   Tetrahedral Electronic Geometry: ABU₃ Species
(Three Lone Pairs of Electrons on A)

The hydrogen halides, HF, HCl, HBr, and HI, are familiar examples. All are gases at normal temperatures and pressures.

Dot Formulas   Electronic Geometry

Molecular Geometry   Polarity

8-11   Trigonal Bipyramidal Electronic Geometry: AB₅,
AB₄U, AB₃U₂, and AB₂U₃

1.   Two examples of AB₅ molecules are PF₅ and AsF₅. Both are *trigonal bipyramidal, nonpolar molecules*. Consider AsF₅.

Electronic Structures of Atoms   Dot Formula

Electronic Geometry about As   VSEPR

Molecular Geometry   Polarity

Valence Bond Theory (Hybridization)

Some modifications of AB_5 molecules include:

1. SF_4, an AB_4U molecule

VSEPR   Molecular Geometry

2. IF_3, an AB_3U_2 molecule

VSEPR   Molecular Geometry

3. XeF_2, an AB_2U_3 molecule

VSEPR   Molecular Geometry

8-12 Octahedral Electronic Geometry: AB_6, AB_5U, and AB_4U_2

Two examples of AB_6 molecules are SF_6 and SeF_6. Both are *octahedral, nonpolar* molecules. Consider SeF_6.

Electronic Structures of Atoms

Dot Formula

Electronic Geometry about Se

VSEPR

Molecular Geometry

Polarity

Valence Bond Theory (Hybridization)

Some modifications of AB_6 molecules include:

1. IF_5, an AB_5U molecule

VSEPR

Molecular Geometry

128

2. XeF_4, an AB_4U_2 molecule

<u>VSEPR</u>   <u>Molecular Geometry</u>

We may summarize some of our observations on simple covalent molecules.

1.   The *electronic geometry* (around the central atom) is determined from the number of regions of high electron density around the central atom shown in the dot formulas.

2.   Sets of electron pairs achieve maximum separation around the central atom.

3.   The *molecular geometry* is determined by locations of *atoms* around the central atom.

4.   Molecules that have symmetrical distributions of electron density are nonpolar molecules.

5.   Molecules that have unsymmetrical distributions of electron density are polar molecules.

6.   Hybridization is determined by *electronic* geometry.

You should study Tables 8-1, 8-2, 8-3, and 8-4 very thoroughly!

8-13   Compounds Containing Double Bonds

In Chapter 7 we drew dot formulas for some species that contain double bonds. The best examples of compounds that contain double bonds are the olefinic hydrocarbons that include $H_2C=CH_2$, $H_3C-CH=CH_2$, and so on. Ethene (ethylene), $H_2C=CH_2$, is the simplest example. Its dot formula is

Because there are the three regions of high electron density around each carbon atom, VSEPR theory suggests that each carbon atom is at the center of an equilateral triangle.

129

Valence Bond Theory postulates sp^2 hybridization for the carbon atoms with one electron remaining in an unhybridized $2p$ orbital, which is perpendicular to the plane of the three sp^2 hybrid orbitals. Top and side views (See Figure 8-4) of these orbitals may be represented as

The C = C bond results from the *head-on* overlap of two sp^2 hybrid orbitals. A *sigma* bond is formed. The side-on overlap of the $2p$ orbitals on adjacent C atoms forms a *pi* bond. See Figures 8-5 and 8-6.

The four atoms attached to the doubly bonded carbon atoms lie in the same plane. In larger molecules, such as $H_3C-CH=CH_2$, each double bonded carbon atom lies at the center of an equilateral triangle. The four atoms bonded to these carbon atoms lie in the same plane (at the corners of the two co-planar equilateral triangles that share a common corner). The carbon atom in the H_3C- group lies at the center of a tetrahedron.

8-14   Compounds Containing Triple Bonds

Ethyne (acetylene), HC≡CH, is the simplest hydrocarbon that contains a triple bond. Its dot formula is

Because there are two regions of high electron density around each carbon atom, VSEPR theory predicts that they are 180° apart.

Valence Bond theory suggests sp hybridization at each carbon atom, with one electron in each of the two unhybridized $2p$ orbitals on each C atom. These are 90° apart and 90° from the sp hybrid orbitals on each carbon atom. See Figure 8-7.

The *head-on* overlap of the two half-filled $2p$ orbitals on each carbon atom results in the formation of a *sigma* bond between the two carbon atoms.

The *side-on* overlap of the two half-filled $2p$ orbitals on each carbon atom results in the formation of two *pi* bonds. See Figure 8-8.

8-15 A Summary of Electronic and Molecular Geometries

Table 8-4 summarizes our discussion to this point. This table includes examples of molecules and polyatomic ions. Study it carefully.

Synthesis Question

The basic shapes that we have discussed in Chapter 8 are present in most molecules. The chemical structure of vitamin B_6 phosphate is shown below. What are the electronic geometry and hybridization of each of the indicated atoms in vitamin B_6 phosphate?
There is a C atom at each of the five other corners of the hexagonal part of the structure.

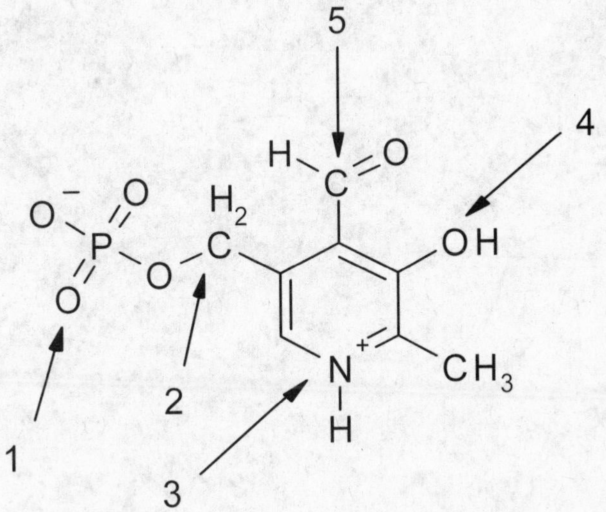

Group Question

The structure of penicillin-G is shown below. What are the electronic geometry and hybridization of each of the indicated atoms in penicillin-G?

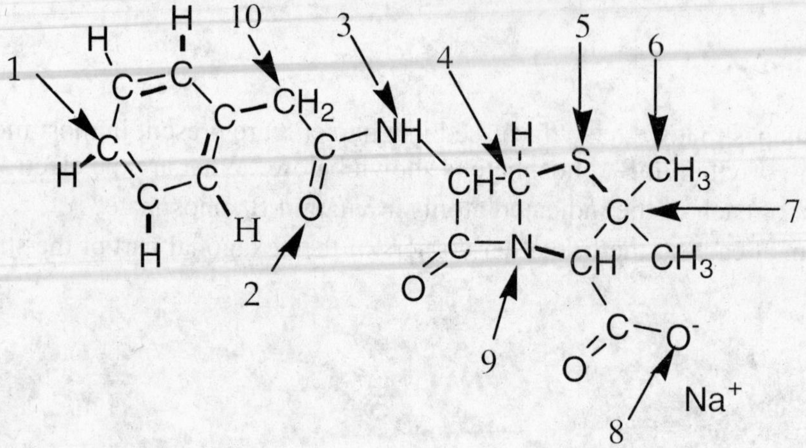

Chapter Nine

MOLECULAR ORBITALS IN CHEMICAL BONDING

According to Valence Bond theory, covalent bonds result from the sharing of electrons in overlapping (hybridized or unhybridized) atomic orbitals localized on different atoms. In molecular orbital (MO) theory, we postulate the combination of atomic orbitals *on different atoms to form molecular orbitals (MO's) so that electrons in them belong to the molecule as a whole*. The two theories are complementary ways of looking at covalent bonding.

9-1 Molecular Orbitals

The electron waves that describe atomic orbitals have positive and negative phases or amplitudes. When they are combined (as in hybridization, according to VB theory) in the formation of MO's they may interact either constructively or destructively. Consider (a) addition and (b) subtraction of the waves shown below.

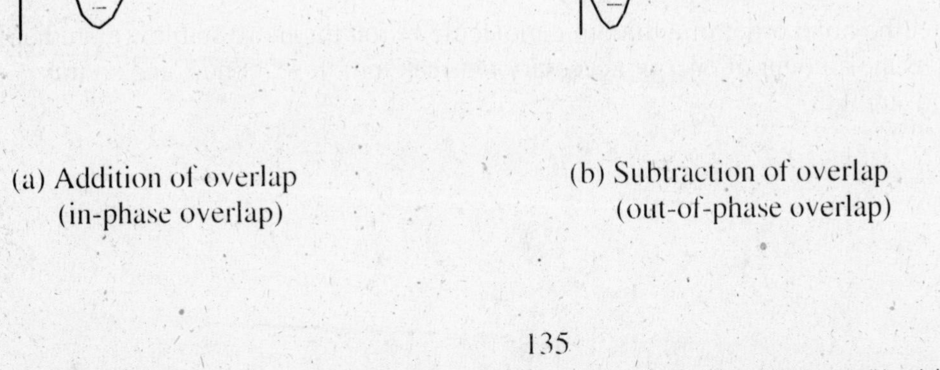

(a) Addition of overlap
(in-phase overlap)

(b) Subtraction of overlap
(out-of-phase overlap)

Let us now describe the kinds of molecular orbitals that can be produced by overlap of atomic orbitals. *Sigma,* (σ) molecular orbitals result from the *head-on* overlap of atomic orbitals; *pi (π)* MO's result from the *side-on* overlap of orbitals. The overlap of two *s* atomic orbitals produces a σ_S MO and a σ_S^* MO. See Figure 9-2.

The σ_S orbital is lower in energy than the combining *s* orbitals and is called a *bonding orbital*. The σ_S^* orbital is higher in energy than the *s* orbitals and is called an *antibonding orbital* (designated by *). The total electron capacity of a set of MO's is the same as that of the atomic orbitals from which they are formed.

All sigma antibonding orbitals have nodal planes that bisect the internuclear axis. A *node* or a *nodal plane* is a region in which the probability of finding electrons is zero.

The head-on overlap of two corresponding *p* orbitals on different atoms (say p_x with p_x) yields σ_p and σ_p^* molecular orbitals. See Figure 9-3.

The side-on overlap of two corresponding *p* orbitals on different atoms (say p_y with p_y or p_z with p_z) produces π and π^* MO's. See Figure 9-4.

9-2   Molecular Orbital Energy Level Diagrams

The order of filling of MO's is governed by the same rules as atomic orbitals. See Figure 9-5. The Aufbau Principle and Hund's Rule are obeyed.

9-3   Bond Order and Bond Stability

The *bond order* (bo) of a molecule is defined as half the number of electrons in bonding orbitals minus half the number of electrons in antibonding orbitals.

The greater the bond order of a diatomic molecule or ion the more stable we predict it to be. Bond energy is the amount of energy necessary to break a mole of bonds, and so it is measured by bond strength.

9-4   Homonuclear Diatomic Molecules

Consider the overlap of the atomic orbitals of two nitrogen atoms to form an N_2 molecule. Each N atom has 7 electrons and therefore an N_2 molecule has 14 electrons.

In shorthand notation we represent this configuration as

The greater the bond order of a bond the more stable we predict it to be. For N_2 the bond order is

This corresponds to a triple bond from the Valence Bond viewpoint. The bond energy is correspondingly high, 946 kJ/mol. N_2 is a very stable molecule. N_2 is quite unreactive; this property is attributed to the high bond energy.

The MO treatment can also be applied to ions. However, ions possess charges and charge as well as bond order influences stability.

Example 9-1:   Write out the electron configuration of the N_2^+ ion in abbreviated notation $(\sigma_{1s}^2 \sigma_{1s}^{*2} ...)$. What is the bond order?

Example 9-2:   Write out the electron configuration for the O_2 molecule in abbreviated notation. What is the bond order? Is the molecule paramagnetic or diamagnetic?

Example 9-3:   Write out the electron configuration for the Be_2 molecule in abbreviated form. What is the bond order? Would you predict that the molecule exists?

137

Example 9-4:   Write out the electron configuration for the F_2 molecule. In this molecule the σ_p MO is lower in energy than the π_p MOs. What is the bond order? Is F_2 paramagnetic?

9-5   Heteronuclear Diatomic Molecules

When MOs are drawn for heteronuclear molecules the energies of the combining atomic orbitals are skewed to take into account differences in electronegativities. The atomic orbitals of the more electronegative element are lower in energy than those of the less electronegative element. Consider the hydrogen fluoride molecule, HF. See Figure 9-9.

9-6   Delocalization and Shapes of Molecular Orbitals

In Section 7-9 we described resonance formulas for molecules and polyatomic ions. In MO theory, we speak of *delocalization* of electrons. Consider the carbonate ion, CO_3^{2-}.
See Figure 9-10.

The structure of benzene, C_6H_6, is explained well by MO theory. See Figure 9-11.

Synthesis Question

Three electron density plots of benzene are shown below. In these pictures, regions of high electron density are red colored and regions of low electron density are blue colored. Use these plots to answer these questions.

Is benzene more susceptible to nucleophilic attack (chemical species that are seeking positively charged regions of electron density) or to electrophilic attack (chemical species that are seeking negatively charged regions of electron density)? Which atoms are more likely to react in benzene, C or H?

Group Question

Shown below are three electron density plots of nitrobenzene. Use these plots to answer these questions.

Is nitrobenzene more susceptible to nucleophilic attack (chemical species seeking positively charged regions of electron density) or electrophilic attack (chemical species seeking negatively charged regions of electron density)?

Which atoms are most likely to react, C, H, N or O?

Chapter Ten

REACTIONS IN AQUEOUS SOLUTIONS I:
ACIDS, BASES, AND SALTS

Many acids, bases, and salts are very important compounds because they occur in living systems, where they perform many essential functions. Additionally many acids, bases, and salts are important industrial chemicals and medicines.

10-1   Properties of Aqueous Solutions of Acids and Bases

Aqueous solutions of most protic acids have some properties in common with each other. They (1) taste sour, (2) change the colors of many indicators, (3) react with active metals to liberate hydrogen, $H_{2(g)}$ (oxidizing acids produce other products), (4) react with metal oxides and hydroxides to form salts and water, (5) react with salts of weaker acids to form the weaker acid and the salt of the stronger acid, and (6) their aqueous solutions conduct electricity.

Solutions of bases (1) taste bitter, (2) have a slippery feeling, (3) change the colors of many indicators, (4) react with acids to form salts and water, and (5) their aqueous solutions conduct electricity.

10-2   The Arrhenius Theory

The classical acids and hydroxide bases discussed in Section 4-9 are also called "Arrhenius acids and bases". According to the **Arrhenius theory**, an *acid* contains hydrogen and ionizes in aqueous solutions to produce H^+ ions. A base contains an -OH group and ionizes or *dissociates* in aqueous solutions to produce hydroxide ions, OH^-.

143

Acids

Bases

Neutralization is the combination of hydrogen ions with hydroxide ions to form water. The net ionic equation for reactions of *strong acids* (Table 4-5) with *strong soluble bases* (Table 4-7) to form *soluble salts* and water is:

Net ionic equations for other kinds of acid-base reactions (neutralizations) are different.

10-3   The Hydronium Ion (Hydrated Hydrogen Ion)

Arrhenius described H^+ ions in aqueous solutions as bare protons. We now know that they are hydrated, $H(H_2O)_n^+$, in which n is a small integer. We usually represent the hydrated hydrogen ion as H_3O^+ in which n = 1. This hydronium ion is the species that gives aqueous solution of acids their characteristic acidic properties.

10-4   The Brønsted-Lowry Theory

In the Brønsted-Lowry theory acids and bases are defined as follows.

Acid - a proton (H^+) donor

Base - a proton (H^+) acceptor

An acid-base reaction is the transfer of a proton, H^+, from an acid to a base.

144

Conjugate acid-base pairs are species that differ by a proton. The weaker a molecular acid is, the stronger its anion is as a base. The F⁻ ion is a stronger base than the NO_3^- ion.

The Brønsted-Lowry definitions are much less restrictive than those of Arrhenius. For example, ammonia can be considered a (weak) base when dissolved in water, even though it does *not* contain a -OH group. However, aqueous solutions of ammonia contain much higher concentrations of OH⁻ ions than pure water.

The weaker a base is, the stronger its characteristic cation is as an acid. For example, the NH_4^+ ion is a stronger Brønsted-Lowry acid than Na^+, the cation of a strong soluble base.

Aqueous ammonia and its derivatives, the amines are the common weak bases. In aqueous solutions the amines ionize in reactions similar to those of ammonia.

Water can act as both an acid and a base according to the Brønsted-Lowry. Such species are said to be *amphoteric (amphiprotic*, when the reactions involve transfer of a proton, as *all* Brønsted-Lowry acid-base reactions do).

10-5   The Autoionization of Water

Careful experiments show that the autoionization of water forms small, equal concentrations of H⁺ and OH⁻ ions. In simplified notation we represent this reaction as

We can write this equation in more detail to show the nature of this reaction more fully.

We saw in Section 4-11, Part 1, that H_3O^+ and OH^- ions combine to form nonionized water molecules when strong acids and strong soluble bases react.

10-6   Amphoterism

The hydroxides of elements of intermediate electronegativity, such as zinc and aluminum, are *amphoteric*. That is, they can react with and act as both acids and bases. Consider the reaction of zinc hydroxide with HNO_3, in which $Zn(OH)_2$ acts as a base.

formula unit

total ionic

net ionic

We may represent the reaction in more detail as

$Zn(OH)_2$ can also act as an *acid* when it dissolves (by reaction) with an excess of a strong soluble base such as potassium hydroxide.

formula unit

total ionic

net ionic

We may represent the reaction in more detail as

Common amphoteric oxides and hydroxides are listed in Table 10-1. Study it carefully.

10-7   Strengths of Acids

Binary Acids

The common binary acids are the hydrogen halides, HF, HCl, HBr, and HI, and the hydrides of the Group 6A elements, H_2O, H_2S, H_2Se, and H_2Te. Within each series, acid strength increases with decreasing bond strength. The order of increasing *bond* strengths and *acid* strengths for the hydrogen halides is

For the 6A hydrides, the order is

It is not possible to distinguish among the strengths of acids stronger than H_3O^+ in water, because all these acids are *leveled* to the strength of H_3O^+ by water. For example, for HI in water:

Likewise, for HBr in water

Both HI and HBr are stronger acids than H_3O^+, but their difference in acid strengths cannot be determined in water. Both are very nearly 100% ionized in dilute aqueous solutions. When we use a solvent less basic (more acidic) than water, HI ionizes to a greater extent than HBr. Anhydrous acetic acid is such a solvent. The relative acid and base strengths of a few acids and their conjugate bases are (See Table 10-2).

147

Acid		Conjugate Base
Strongest Acid		Weakest Base
	Acid loses H$^+$	
	$\rightarrow$	
	$\leftarrow$	
	base gains H$^+$	
Weakest Acid		Strongest Base

Hydronium ion, H$_3$O$^+$, is the strongest acid that can exist in aqueous solutions.

Hydroxide ion, OH$^-$, is the strongest base that can exist in aqueous solutions.

Ternary acids

Ternary acids are *hydroxides of nonmetals*, but they ionize to produce H$_3$O$^+$ ions. The structures of perchloric, HClO$_4$, and phosphoric, H$_3$PO$_4$, acids are

Weak ternary acids have stronger H-O bonds than stronger ternary acids.

| acid strength: | HNO_2 | HNO_3 | H_2SO_3 | H_2SO_4 |

| H-O bond strength: | HNO_2 | HNO_3 | H_2SO_3 | H_2SO_4 |

Acid strengths of ternary acids usually increase with

1. Increasing number of O atoms/central nonmetal atom.

2. Increasing oxidation state of central nonmetal atom.

For two (or more) ternary acids involving the *same central element*, the acid with the central element in the highest oxidation state (greater number of oxygen atoms per central atom) is usually the stronger acid. The order of increasing acid strengths of the ternary acids of chlorine is

weakest _____<_____<_____<_____strongest

Acid strengths of most ternary acids that contain *different central elements,* in the *same oxidation state* from the *same group in the periodic table*, increase with increasing electronegativity of the central atom.

10-8   Acids-Base Reactions in Aqueous Solutions

In Section 4-11 we discussed some acid-base reactions. Review that discussion.

Let us now classify acid-base reactions according to the characteristics of the acid and the base that react, as well as of the salt formed by the reaction. *Assume complete neutralization.* The essence of each reaction (net ionic equation) is similar or the same for every reaction in a given category, and different from those in other categories.

A few useful generalizations are: (1) Strong acids and strong soluble bases are predominantly ionized or dissociated in aqueous solutions. (2) Nearly all soluble salts (Sec. 4-2) are dissociated; exceptions will be noted as they are encountered. (3) Weak acids and weak bases exist primarily as nonionized molecules. (4) Insoluble compounds exist as undissociated formula units or nonionized molecules.

Reactions of Strong Acids with Strong Soluble Bases to Form Soluble Salts

Nearly all strong acid react with strong soluble bases to form soluble salts (Section 4-2). For example, HBr reacts with $Ca(OH)_2$ to form $CaBr_2$ and H_2O.

formula unit

total ionic

net ionic

This is the *net ionic equation* for all reactions of strong acids with strong soluble bases that form soluble salts.

Reactions of Weak Acids and Strong Soluble Bases to Form Soluble Salts

Nitrous acid reacts with sodium hydroxide to form sodium nitrite.

formula unit

total ionic

net ionic

In general terms the reactions of weak monoprotic acids with strong soluble bases to form soluble salts may be written as

Reactions of Strong Acids with Weak Bases to Form Soluble Salts

Nitric acid reacts with aqueous ammonia to form ammonium nitrate.

formula unit

total ionic

net ionic

Reactions of Weak Acids with Weak Bases to Form Soluble Salts

Acetic acid reacts with aqueous ammonia to form ammonium acetate.

formula unit

total ionic

net ionic

There are other kinds of acid-base reactions. We have illustrated the basic ideas.

10-9   Acidic Salts and Basic Salts

The reactions of stoichiometric amounts of acids and bases produce *normal salts*. These contain no unreacted H^+ or OH^- ions.

The reaction of a polyprotic acid with less than a stoichiometric amount of base produces an *acidic salt*. These contain one or more acidic H's per formula unit. Consider the reactions of 1:1 and 1:2 mole ratios of sulfuric acid and sodium hydroxide.

1:1

1:2

Basic salts contain one or more unreacted OH^- per formula unit. The reactions of 1:1 and 1:2 mole ratios of barium hydroxide and hydrochloric acid produce a *basic salt* and a *normal salt*.

1:1

1:2

Acidic salts and basic salts are capable of neutralizing bases and acids, respectively. However, solutions of such salts are *not* necessarily acidic or basic.

151

10-10   The Lewis Theory

According to this theory acids, bases, and neutralization are defined very broadly.

Acid   a species that accepts a share in an electron pair.

Base   a species that makes available, or "donate" a share in an electron pair.

Neutralization   the formation of a *coordinate covalent bond*, a bond in which both electrons originated on the same (*donor*) atom.

The following are examples of Lewis acid-base reactions.

Ionization of aqueous NH_3

Ionization of aqueous HBr

Autoionization of water

The reaction of sodium fluoride with boron trifluoride is a Lewis acid-base reaction. It is *not* an acid-base reaction according to the *Arrhenius* and *Brønsted-Lowry* theories.

The reaction of gaseous NH_3 with gaseous HBr is a Lewis *and* a Brønsted-Lowry acid-base reaction, but *not* an Arrhenius acid-base reaction.

The Lewis theory is very broad and it covers many reactions. As our knowledge of chemical reactions increases, the Lewis theory becomes more useful.

10-11　The Preparation of Acids

Binary acids can be prepared by combination of the appropriate elements with hydrogen. Review Section 6-7, part 2.

Volatile acids can be prepared by the treatment of their salts with a nonvolatile acid such as concentrated sulfuric (b.p. 338°C) or phosphoric acid (b.p. 213°C). Concentrated H_2SO_4 is a strong oxidizing agent. H_3PO_4 is used to prepare HBr (b.p. -67°C) and HI (b.p. -35°C), because concentrated H_2SO_4 oxidizes HBr and HI to Br_2 and I_2, respectively.

Many *ternary acids* can be prepared by combination of the appropriate nonmetal oxides (acid anhydrides) with water.

Halides and oxyhalides of some nonmetals react with water to produce (1) a binary hydrohalic acid, and (2) the ternary acid of the other nonmetal.

Synthesis Question

One method of increasing the solubility and the absorption of medications is to convert weakly acidic drugs into sodium salts before making the pills that will be ingested. How does this preparation method enhance the drug's solubility in the stomach?

Group Question

Medicines that are weakly basic are not absorbed well into the bloodstream. One method to increase their absorption is to take an antacid at the same time that the medicine is administered. How does this method increase the absorption?

Chapter Eleven

REACTIONS IN AQUEOUS SOLUTIONS II: CALCULATIONS

AQUEOUS ACID-BASE REACTIONS

11-1   Calculations Involving Molarity

Sometimes we need to calculate the concentration of a solution prepared by mixing solutions of two substances that react with each other. The *reaction ratio* is the relative number of moles of reactants and products shown in the balanced chemical equation.

Example 11-1:   If 100.0 mL of 1.00 M NaOH and 100.0 mL of 0.500 M H_2SO_4 solutions are mixed, what will the concentration of the resulting solution be?

$$2\ NaOH\quad +\quad H_2SO_4\quad \rightarrow\quad Na_2SO_4\quad +\quad 2\ H_2O$$

Reaction ratio:

Before reaction:
_____

After reaction:

Example 11-2:   If 130.0 mL of 1.00 M KOH and 100.0 mL of 0.500 M H_2SO_4 solutions are mixed, what will be the concentrations of KOH and K_2SO_4 in the resulting solution?

$$2\ KOH \quad + \quad H_2SO_4 \quad \rightarrow \quad K_2SO_4 \quad + \quad 2\ H_2O$$

Reaction ratio:

Before reaction: _____

After reaction:

Example 11-3:   What volume of 0.750 M NaOH solution would be required to neutralize completely 100. mL of 0.250 M H_3PO_4 solution?

$$3NaOH \quad + \quad H_3PO_4 \quad \rightarrow \quad Na_3PO_4 \quad + \quad 3H_2O$$

11-2   Titrations

Titration is the process by which one determines the volume of a standard solution required to react with a specific amount of another substance. Both of the concentrations of the solutions are given. See Fig. 11-1.

Standard solutions are solutions of accurately known concentrations.

Standardization is the process by which one determines the concentration of a solution by measuring the volume of the solution required to react with a known amount of primary standard.

An *indicator* is a highly colored substance that can exist in different forms, with different colors that depend on the concentration of H^+ ions in the solution.

The *equivalence point* is the point at which chemically equivalent amounts of acid and base have reacted.

The *end point* is the point at which the indicator changes color and the titration is stopped.

Primary standards are compounds that (1) do not react with the atmosphere (particularly H_2O and CO_2), (2) do not react with the atmosphere, (3) are soluble in water, (4) have high formula weights, (5) react according to one invariable reaction.

11-3   Calculations for Acid-Base Titrations

Standard solutions of bases are not easily prepared directly because bases react with both CO_2 and H_2O in the atmosphere. So weighing out a sample of a base accurately is difficult. Solutions of bases of the approximately desired concentration are prepared. These solutions are then standardized with a primary standard.

The common primary standard used to standardize solutions of bases is potassium hydrogen phthalate (KHP). The structure of potassium hydrogen phthalate is

Each formula unit of KHP contains one acidic hydrogen. Therefore one mole (204.2 g) of KHP can react with one mole of OH^- ions.

Example 11-4:   Calculate the molarity of a sodium hydroxide solution if 27.3 mL of the solution reacts with 0.4084 gram of KHP.

NaOH   +   KHP $\rightarrow$ NaKP   + H_2O

Sodium carbonate (Na_2CO_3) is a common primary standard used to standardize solutions of acids. It reacts with acids as shown below.

$$Na_2CO_3 \quad + \quad 2HCl \quad \rightarrow \quad 2NaCl \quad + \quad CO_2(g) \quad + \quad H_2O$$

$$Na_2CO_3 \quad + \quad H_2SO_4 \quad \rightarrow \quad Na_2SO_4 \quad + \quad CO_2(g) \quad + \quad H_2O$$

Example 11-5: Calculate the molarity of a sulfuric acid solution if 23.2 mL of it reacts with 0.212 grams of sodium carbonate.

$$Na_2CO_3 \quad + \quad H_2SO_4 \quad \rightarrow \quad Na_2SO_4 \quad + \quad CO_2(g) \quad + \quad H_2O$$

Example 11-6: An impure sample of potassium hydrogen phthalate, KHP, had a mass of 0.884 g. It was dissolved in water and titrated with 31.5 mL of 0.100 M NaOH solution. Calculate the percent purity of the KHP sample. Molar mass of KHP is 204.2 g/mol.

$$NaOH \quad + \quad KHP \quad \rightarrow \quad NaKP \quad + \quad H_2O$$

The *equivalent weight* of a *base* is the mass of the base, expressed in grams, that furnishes 6.022×10^{23} OH^- ions, or that reacts with 6.022×10^{23} H_3O^+ ions.

The *equivalent weight* of an *acid* is the mass of the acid, expressed in grams, that furnishes 6.022×10^{23} H^+ (H_3O^+) ions, or that reacts with 6.022×10^{23} OH^- ions. One mole of an acid contains 6.022×10^{23} formula units.

160

OXIDATION-REDUCTION REACTIONS

All balanced equations must satisfy the following two criteria.

1. There has to be a mass balance. The same number of atoms of each kind must appear in a reactants and products.

2. There must be a charge balance. The sums of the charges on the left and right sides of the equation must equal each other.

11-4   Balancing Redox Equations

In the half-reaction method, redox reactions are separated into **half-reactions**, one representing oxidation and the other reduction. There is a straight-forward procedure for balancing the half-reaction.

1. Write as much of the overall unbalanced equation as possible, omitting spector ions.

2. Construct unbalanced oxidation and reduction half-reactions (these are usually incomplete as well as unbalanced). Show complete formulas for polyatomic ions and molecules.

3. Balance by inspection all elements in each half-reaction, except H and O. Then use the rules from Section 11-6 to balance H and O in each half-reaction.

4. Balance the charge in each half-reaction by adding electrons as "products" or "reactants."

5. Balance the electron transfer by multiplying the balanced half-reactions by appropriate integers.

6. Add the resulting half-reactions and eliminate any common terms.

7. Add common species that appear on the same side of the equation and cancel equal amounts of common species that appear on opposite sides of the equation in equal amount.

8. Check for mass and charge balance.

Example 11-7:   Tin (II) ions are oxidized to tin (IV) ions by bromine. Use the half-reaction method to write and balance the net ionic equation.

oxidation

reduction

11-5   Adding H⁺, OH⁻, or H₂O to Balance Hydrogen or Oxygen

Often more oxygen or hydrogen is needed to complete the mass balance for a redox reaction in aqueous solution.

For redox reactions that occur

a)   in acidic solution, H^+ or H_2O may be added to either side of the equation, as needed, to give mass and charge balance.

b)   in basic solutions, OH^- or H_2O may be added as necessary.

Every BALANCED equation has mass AND charge balance.

Example 11-8:   Dichromate ions oxidize iron (II) ions to iron (III) ions and are reduced to chromium (III) ions in *acidic* solution. Write and balance the net ionic equation for the reaction.

oxidation

reduction

162

Example 11-9:   In basic solution hydrogen peroxide oxidizes chromite ions, $Cr(OH)_4^-$, to chromate ions, CrO_4^{2-}. Write and balance the net ionic equation for the reaction.

oxidation

reduction

Example 11-10:   When chlorine is bubbled into basic solution, it forms hypochlorite ions and chloride ions. Write and balance the net ionic equation.

oxidation

reduction

11-6   Calcuations for Redox Reactions

Redox titrations are often used to determine the amount of an oxidizable or a reducible substance. In such a titration we determine the volume of a reagent of known concentration that reacts with a known amount of the other reactant. $KMnO_4$ solutions are often used as oxidizing reagents. The reduction of MnO_4^- ions produces significant color changes so the change of color serves to indicate the endpoint of the titration.

Example 11-11:   What volume of 0.200 M $KMnO_4$ is required to oxidize 35.0 mL of 0.150 M HCl?

$$2KMnO_4 + 16HCl \rightarrow 2KCl + 2MnCl_2 + 5Cl_2 + 8H_2O$$

Example 11-12:   A volume of 40.0 mL of iron (II) sulfate is oxidized to iron (III) by 20.0 mL of 0.100 M potassium dichromate solution. What is the concentration of the iron (II) sulfate solution? See Example 11-14 for the reaction.

Synthesis Question

A 0.7500 g sample of an impure $FeSO_4$ sample is titrated with 26.25 mL of 0.02000 M $KMnO_4$ to an endpoint. What is the % purity of the $FeSO_4$ sample?

Group Question

Ice cubes that are made from water that is found in limestone rich areas of the country, such as Florida, have an unusual property. When a soft drink, like Coca-Cola, is poured over them an excessive amount of "fizzing" occurs. Write the molecular, total, and net ionic equations for the reaction that causes the excessive fizzing.

Chapter Twelve

GASES AND THE KINETIC-MOLECULAR THEORY

12-1 Comparison of Solids, Liquids, and Gases

Matter is arbitrarily classified into three physical states: solids, liquids, and gases. Most substances can exist in the gas phase, and most can be liquefied and then solidified at appropriate temperatures and pressures. The volumes of gases change greatly as temperature and pressure change. Volumes of solids and liquids change very little with changes in temperature and pressure. Gases are less dense than solids and liquids (see Table 12-1).

12-2 Composition of the Atmosphere and Some Common Properties of Gases

The atmosphere is a mixture of gases (Table 12-2). Some observations on gases are:

1. Gases can be compressed.

2. Gases exert pressure on their surroundings.

3. Gases can expand without limit.

4. Gases diffuse into each other, i.e., they are miscible (mix with each other) in all proportions unless they react chemically.

5. Gases can be described in terms of their volume, temperature, pressure, and the number of molecules (moles) of gas present.

12-3   Pressure

Pressure is defined as *force per unit area*. It may be expressed in many units. Average atmospheric pressure at sea level is 14.7 pounds per square inch.
The mercury barometer is a convenient device for measuring pressure. The *average pressure at sea level* is referred to as *standard pressure*. Standard pressure is 76 cm of mercury, 760 mm of mercury or *760 torr*, and is referred to as *one atmosphere* of pressure. In the SI system of units, one atmosphere is 1.013×10^5 pascals (Pa) or 101.3 kilopascals (kPa).

12-4   Boyle's Law: The Pressure-Volume Relationship

Boyle's Law is usually stated: *At constant temperature, the volume occupied by a definite mass of a gas is inversely proportional to the pressure applied to the gas.* This statement is a summary of many observations on the behavior of gases. See Figure 12-3.

Example 12-1:   At 25°C a sample of helium occupies 400 mL under a pressure of 760 torr. What volume would it occupy under a pressure of 2.00 atmospheres at the same temperature?

12-5   Charles' Law: The Volume-Temperature Relationship; The Absolute Temperature Scale

The fact that gases expand when heated and contract when cooled has been common knowledge for many years. However, evolution of a simple relationship between the volume and temperature of a sample of gas had to await the evolution of an absolute temperature scale. Charles and Gay-Lussac studied the relationship between the volume and temperature of a sample of gas at constant pressure. Charles observed that a plot of volume versus temperature gives a straight line. Charles' Law is usually stated: *at constant pressure the volume occupied by a definite mass of a gas is directly proportional to the absolute temperature.*

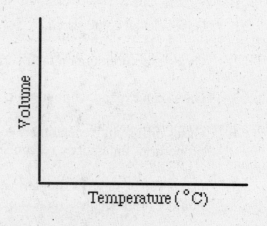

168

The volume-temperature line intercepts the temperature axis at -273.15°C. This temperature is taken as absolute zero or 0K. The relationship between the Celsius (centigrade) and the absolute (Kelvin) temperature scales is

Charles' observations may be represented symbollically as

Example 12-2:   A sample of hydrogen, H_2, occupies 100 mL at 25°C and 1.00 atmosphere. What volume would it occupy at 50°C under the same pressure?

12-6   Standard Temperature and Pressure

Because volumes of gases vary with both temperature and pressure, we establish reference points for temperature and pressure. These are called *standard conditions* (SC or STP) and refer to 0°C (273.15K) and *one atmosphere of pressure* (760 torr or 101.3 kPa).

12-7   The Combined Gas Law Equation

Boyle's Law tells us that at constant temperature the volume occupied by a definite mass of a gas is inversely proportional to the pressure under which it is confined. Charles' Law tells us that at constant pressure the volume occupied by a definite mass of the gas is directly proportional to the absolute temperature.

For a given sample of gas, the relationship among temperature, volume, and pressure is

Or, in a more useful form we write the *combined gas law equation* as

This equation contains six variables. If five are known, we can find the sixth.

169

Example 12-3:   A sample of nitrogen, N_2, occupies 750. mL at 75°C under a pressure of 810. torr. What volume would it occupy at standard conditions?

Example 12-4:   A sample of methane, CH_4, occupies 260. mL at 32°C under a pressure of 0.500 atmosphere. At what temperature would it occupy 500. mL under a pressure of 1200. torr?

12-8   Avogadro's Law and The Standard Molar Volume

Avogadro's Law states that *at the same temperature and pressure, equal volumes of all gases contain the same number of molecules (moles).*

The volume occupied by one mole of a gas at standard conditions is *the standard molar volume.* One mole of an ideal gas occupies 22.4 liters at 0°C and one atmosphere of pressure. Real gases do not behave exactly as ideal gases. However, at reasonable temperatures and pressures the behavior of many gases is nearly ideal. We may assume ideal behavior without introducing serious errors into our calculations. The *ideal* standard molar volume, 22.4 liters, is a good approximation of the standard molar volume of many real gases (Table 12-3).

Example 12-5:   One (1.00) mole of a gas occupies 36.5 liters, and its density is 1.36 g/L at given temperature and pressure. (a) What is its molecular weight? (b) What is the density of the gas at standard conditions?

(a)

(b)

12-9   Summary of the Gas Laws : The Ideal Gas Equation

The relationship among the pressure, volume, temperature, and the number of moles in a sample of gas is known as *the ideal gas equation*.

<u>Boyle's Law</u>   <u>Charles' Law</u>   <u>Avogadro's Law</u>

Ideal Gas Equation:

The proportionality constant, R, is the *universal gas constant*, a "constant of nature." One mole of an ideal gas occupies 22.4 liters at standard conditions. Substituting these values into the ideal gas equation, $PV = nRT$, gives the value of R.

The units of R depend on the units used for P, V, and T in the ideal gas equation. Because temperature must be expressed on an absolute temperature scale, we express temperature in kelvins (K). Volumes are usually expressed in liters and pressures in atmospheres.

Example 12-6:   What volume would 50.0 g of ethane, C_2H_6, occupy at 140°C under a pressure of 1820 torr?

Example 12-7:   Calculate the number of moles in, and the mass of, an 8.96-liter sample of methane, CH_4, measured at standard conditions.

Example 12-8: Calculate the pressure exerted by 50.0 g of ethane, C_2H_6, in a 25.0 liter container at 25°C.

12-10 Determination of Molecular Weights and Molecular Formulas of Gaseous Substances

In Sections 2-9 we distinguished between the simplest formulas and the molecular (true) formulas of compounds. If the mass of a volume of gas is known, we can use this information *and* an elemental analysis to determine the molecular formula of a compound.

Example 12-9: A compound that contains only carbon and hydrogen is 80.0% C and 20.0% H by mass. At STP 546 mL of the gas has a mass of 0.732 g. What is the molecular (true) formula for the compound?

Example 12-10: A 1.74 g sample of a compound that contains only carbon and hydrogen contains 1.44 g of C and 0.300 g of H. At STP 101 milliliters of the gas has a mass of 0.262 gram. What is its molecular (true) formula?

12-11 Dalton's Law of Partial Pressures

In a mixture of gases each gas exerts the pressure that it would exert if it occupied the volume alone. See Figure 12-6. Dalton's Law is stated: *The total pressure exerted by a mixture of ideal gases is the sum of the partial pressures of those gases.*

172

Example 12-11:   If 100. mL of hydrogen, measured at 25°C and 3.00 atmospheres pressure, and 100. mL of oxygen, measured at 25°C and 2.00 atmosphere pressure were forced into a single container at 25°C, what would be the pressure of the gas mixture?

The mole fraction, X_A, of component A in a mixture in a dimensionless quantity and can be defined as

In a gaseous mixture, we can relate the mole fraction of each component to its partial pressure.

$$X_A = \frac{P_A}{P_{total}} \qquad \text{or} \qquad P_A = X_A \times P_{total}$$

Example 12-12:   The mole fraction of oxygen in the atmosphere is 0.294. Calcuate the partial pressure of O_2 in air when the atomospheric pressure is 1.05 atm.

Vapor Pressure

The vapor pressure of a liquid is the pressure exerted by its gaseous molecules in equilibrium with the liquid.

All liquids exhibit characteristic vapor pressures that increase with temperature. The illustration describes just what vapor pressure is. The following table lists the vapor pressure of water at temperatures near room temperature.

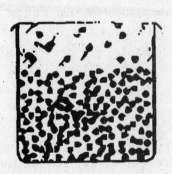

Vapor Pressure (V.P.) of Water Near Room Temperature			
Temp. (°C)	V.P. of Water (torr)	Temp. (°C)	V.P. of Water (torr)
18	15.48	23	21.07
19	16.48	24	22.38
20	17.54	25	23.76
21	18.65	26	25.21
22	19.83	27	26.74

173

Example 12-12:   A sample of hydrogen was collected by displacement of water at 25°C. The atmospheric pressure was 748 torr. What pressure would the dry hydrogen exert in the same volume container at the same temperature.

Example 12-13:   A sample of oxygen was collected by displacement of water. The oxygen occupied 742 mL at 27°C. The barometric pressure was 753 torr. What volume would the dry oxygen occupy at STP?

12-12   Mass-Volume Relationships in Reactions Involving Gases

Oxygen may be prepared in the laboratory by heating potassium chlorate, $KClO_3$, in the presence of a catalyst, MnO_2. One mole of an ideal gas occupies 22.4 liters at standard conditions, so we can also represent the relative amounts of gaseous reactants and products in terms of the standard molar volume.

$$2KClO_3(s) \xrightarrow[\Delta]{MnO_2} 2KCl(s) \quad + \quad 3O_2(g)$$

Three moles of oxygen, 96 g of oxygen, or 67.2 liters of oxygen (measured at standard conditions) represent the same amount of oxygen. We can construct unit factors using any two chemically equivalent quantities from the balanced equation.

Example 12-14:   What volume of oxygen, measured at STP, can be produced by the thermal decomposition of 120. g of $KClO_3$?

Gay-Lussac's Law of Combining Volumes states at *constant temperature and pressure, the volumes of reacting gases can be expressed as a ratio of simple whole numbers.*

12-13   The Kinetic-Molecular Theory

The basic assumptions of the *Kinetic-Molecular theory* follow.

1. Gases consist of discrete molecules. The individual molecules are very small and are very far apart relative to their own sizes.

2. Gaseous molecules are in continuous, rapid random, straight-line motion with varying velocities.

3. The collisions between gas molecules and with the walls of the container are elastic; the total energy is conserved during a collision.

4. Between collisions, the molecules exert no attractive or repulsive forces on one another; each molecule travels in a straight line with a constant velocity.

The average kinetic energy is the energy a body possesses by virtue of its motion. The kinetic energy of a body is $KE = 1/2(mv^2)$, where m is its mass and v is its velocity. The average kinetic energies of molecules of different gases are equal at a given temperature.

Kinetic energies of gases increase with temperature and decrease as temperature decreases. Light molecules such as H_2 and He have much higher velocities than heavier molecules such as CO_2 and SO_2 at a given temperature.

Kinetic-Molecular Theory satisfactorily explains most of the observed behavior of gases. Many experiments have indicated its validity. See Figures 12-10, 12-11, and 12-12.

Boyle's Law

Dalton's Law

Charles's Law

12-14   Diffusion and Effusion of Gases

A sample of gas fills any closed container in which it is placed. When gases are placed in containers with porous walls they *effuse* through the walls of the containers. Lighter gases effuse through the small openings in porous walls more rapidly than heavier gases. Diffusion is the movement of a substance into a space or the mixing of one substance with another. See Figure 12-13 and 12-14.

12-15   Deviations From Ideal Gas Behavior

At ordinary temperatures and pressures most gases obey the gas laws reasonably well. However, at very high pressures or very low temperatures, when the gaseous molecules are forced quite close together, significant deviations from ideal behavior are observed. These deviations are due to the fact that when gaseous molecules are forced very close together two complicating factors must be considered: (1) the volume occupied by the gaseous molecules themselves, and (2) attractions among the molecules may become significant. Various attempts have been made to describe the behavior of gases by "equations of state". Perhaps the best known of these is the *van der Waals equation*. It attempts to correct for the attractive forces between (among) gaseous molecules and for the volume of the gaseous molecules themselves. The van der Waals equation is

The "a" and "b" terms are called van der Waals constants. The "a" is a constant that corrects for the attractions among molecules. The "b" term corrects for the volume of the gas molecules themselves. When the volume of a gas is large (low pressure and/or high temperature), both "a" and "b" are relatively *very* small and may be disregarded. Then the van der Waals equation reduces to the ideal gas law, $PV = nRT$.

Let's examine the attractive forces that exist among gaseous molecules. For nonpolar gases such as the noble gases, He, Ne, Ar, ..., the possibilities for attractive interactions among molecules are rather limited.

These attractions are known as van der Waals attractions *or* London forces.

For *polar* molecules the possibilities of interactions among molecules are greater. Consider water vapor, H_2O, as an example.

Thus, it is easy to see why steam should deviate significantly from ideal behavior at high pressures and low temperatures, i.e., near its liquefaction point.

176

van der Waals constants for some common gases follow:

Gas	a $(L^2 \cdot atm/mol^2)$	b (L/mol)
H_2	0.244	0.0266
He	0.034	0.0237
O_2	1.36	0.0318
Cl_2	6.49	0.0562
NH_3	4.17	0.0371
H_2O	5.46	0.0305

Example 12-19: Calculate the pressure exerted by 84.0 g of ammonia, NH_3, in a 5.00 L container at 200°C using the ideal gas law.

Example 12-20: Solve Example 12-19 using the van der Waals equation.

The pressure calculated in Example 12-20 is _____% higher than in Example 12-19.

177

Synthesis Question

The lethal dose for hydrogen sulfide is 6 ppm. In other words, if in 1 million molecules of air there are six hydrogen sulfide molecules then that air would be deadly to breathe. How many hydrogen sulfide molecules would be required to reach the lethal level in a room that is 77 feet long, 62 feet wide, and 50. feet tall at 1.0 atm and 25.0° C?

Group Question

Tires on cars are typically filled to a pressure of 35 psi at 300K. A tire is 16 inches in radius and 8.0 inches in thickness. The wheel that this tire is mounted on is 8.0 inches in radius. What is the mass of the air in this tire?

Chapter Thirteen

LIQUIDS AND SOLIDS

The solid and liquid states are called *condensed states* because the individual particles (atoms, ions, or molecules) are in close contact with each other and so they interact much more strongly

13-1   Kinetic - Molecular Description of Liquids and Solids

The properties of liquids and solids can be described in terms of the Kinetic-Molecular Theory. See Figure 13-1 and Table 13-1.

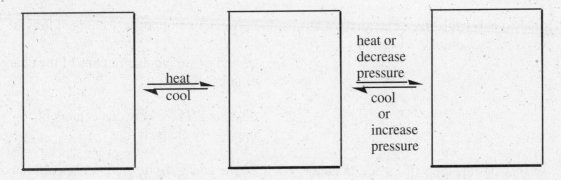

Cooling a gas sufficiently liquefies it as the forces of attraction become greater than the repulsive forces among the individual particles. Further cooling causes the liquid to solidify. The *strengths of interactions* among individual particles *and* the *degree of ordering* of particles increase in the order

In liquids the forces of attraction among particles are great enough to cause disordered clustering. The particles have sufficient energy of motion to overcome partially the attractive forces among them. The particles can slide past each other, and so liquids assume the shapes of their containers.

Miscible liquids diffuse into each other, i.e., they are soluble in each other and form homogeneous solutions.

Immiscible liquids do not diffuse into each other. They are insoluble in each other.

Cooling a liquid lowers its molecular kinetic energy. This causes its molecules to slow down. At sufficiently low temperatures, short-range attractive forces overcome the decreasing kinetic energies of the molecules, and so the liquid solidifies.

Solids consist of particles with a very restricted range of motion. They are held in fixed positions and vibrate about these fixed positions. So solids have definite shapes and volumes. Particles of solids diffuse into each other *very* slowly. See Figure 13-2.

13-2   Intermolecular Attractions and Phase Changes

*Inter*molecular forces are the forces between (among) individual particles (atoms, ions, molecules). They are quite weak relative to *intra*molecular forces, i.e., covalent and ionic bonding *within* compounds. Properties of liquids and solids are strongly dependent on the strengths of *inter*molecular *attractive forces*.

Intermolecular forces are responsible for the existence of solids and liquids. Gases would not condense to liquids and liquids would not solidify if it were not for the existence of intermolecular forces of attraction. We can broadly classify them into four categories. These kinds of interactions are present in all three physical states. They become increasingly significant in the order

gases < liquids < solids.

1. **Ion-ion interactions** - The force of attraction between two oppositely charged ions is governed by Coulomb's Law.

    q^+ and q^- are the magnitudes of the charges on ions

    d is the distance between centers of oppositely charged ions

Energy has the units of force x distance, F x d, so the *energy of attraction* between ions is

Solid state *ionic* compounds consist of extended arrays of ions. Ion-ion interactions can be thought of as both *inter-* and *intra*molecular. Because ionic bonding is quite strong, ionic compounds have high melting points. Compounds containing highly-charged ions have higher melting points than compounds containing univalent ions.

In molten (liquid) ionic compounds, ions are usually further apart than in the solids. Because of the strong attractive forces among ions, ionic compounds, have high boiling points.

182

Example 13-1:   Arrange the following ionic compounds in the expected order of increasing melting and boiling points. NaF, CaO, CaF$_2$

2.   **Dipole - dipole interactions** - Permanent dipole-dipole interactions occur between the δ+ end of one polar molecule and the δ- end of another. They are approximately 1% as strong as typical covalent bonds. (See Figure 13-3.)

Electrostatic forces between ions decrease by the factor 1/d^2 as their separation, d, increases. Dipole-dipole forces vary as 1/d^4, and so they are effective *only* over *very* short distances.

3.   **Hydrogen bonding** - This kind of interaction is an especially strong kind of dipole-dipole interaction. It occurs in polar covalent molecules that contain **H** bonded to one of the very electronegative elements **O, N, or F**. Hydrogen bonding may be up to 10% as strong as typical covalent bonding. See Figure 13-4.

Hydrogen bonding is responsible for the abnormally high melting and boiling points of compounds such as HF, H$_2$O, and NH$_3$. See Figure 13-5.

4.   **Dispersion forces** - These weak interactions are present in all substances. They are effective *only* over *extremely* short distances because they vary as 1/d^7. These are the *only* kinds of attractive interactions *for nonpolar substances*. They result from the attraction of the nucleus of one atom for the electron cloud of another atom (in a different molecule). See Figure 13-6. They increase with increasing *polarizability*, the ability of an electron cloud to be distorted. Polarizability increases with increasing numbers of electrons and therefore with increasing sizes of molecules. See Figure 13-6.

Table 13-3 compares the magnitude of various contributions to the total energy of interaction among molecules for a few substances. Study it carefully.

183

THE LIQUID STATE

The variations in properties of liquids depend upon the nature and strengths of the attractive forces among the particles of a liquid. Some properties of liquids follow.

13-3   Viscosity

Viscosity, resistance to flow, is one measure of the forces of attraction within a liquid. Honey and glycerin are viscous liquids, i.e., they have high viscosities. Gasoline has a low viscosity. See Figure 13-7.

13-4   Surface Tension

Surface tension is a measure of the inward forces that must be overcome to expand the surface of a liquid. Molecules on the surface are attracted only toward the interior, while those in the interior are attracted equally in all directions. See Figure 13-8.

Droplets of water are spherical because the spherical shape minimizes surfaces area.

13-5   Capillary Action

Cohesive forces are the forces that hold a liquid together. *Adhesive forces* are forces between a liquid and another surface. The shape of a meniscus (surface) of a liquid in a small-diameter tube depends upon which forces are greater. Consider glass tubes containing water and mercury. Water is said to "wet" glass, whereas mercury does not. See Figure 13-9.

water   mercury

Water is drawn up through the roots of plants by *capillary action*. This is due to the fact that the adhesive forces between the water and the roots exceed the cohesive forces in water.

13-6   Evaporation

Evaporation (vaporization) is the process in which molecules escape from the surface of a liquid (vaporize). It occurs more rapidly as temperature increases because a greater fraction of molecules possess the necessary escape velocity and kinetic energy. See Figures 13-10 and 13-11.

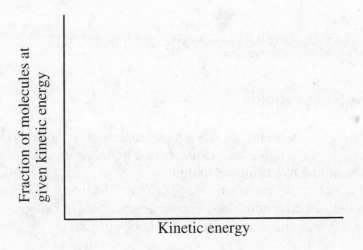

A liquid can eventually evaporate entirely if it is left uncovered and illustrates LeChatelier's Principle which states *a system at equilibrium, or changing toward equilibrium, responds in the way that tends to relieve or "undo" andy stress placed on it.*

13-7   Vapor Pressure

The vapor pressure of a liquid is the pressure exerted by the vapor of the liquid on its surface *at equilibrium* in a closed container. Because vapor pressures increase as temperature increases, evaporation occurs more rapidly as temperature increases. Generally liquids with high boiling points have low vapor pressures and relatively strong intermolecular attractions. See Figures 13-12 and 13-13.

	Vapor Pressures for Three Common Liquids			Normal BP
	0°C	25°C	50°C	
diethyl ether	185 torr	470 torr	1325 torr	_____
methanol	29.7 torr	122 torr	404 torr	_____
water	4.6 torr	23.8 torr	92.5 torr	_____

185

13-8   Boiling Points and Distillation

The boiling point of a liquid is the temperature at which the vapor pressure of the liquid equals the applied (usually atmospheric) pressure. The *normal boiling point* of a liquid is the temperature at which its vapor pressure equals 760 torr (1 atmosphere). See Table 13-5 and Figures 13-12 and 13-13.

Distillation is the process by which a mixture or solution is separated into its components on the basis of differences in boiling points of the components. See Figure 13-14.

13-9   Heat Transfer Involving Liquids

When heat is added to a liquid at a constant rate, its temperature rises (up to its boiling point). The specific heat (Section 1-13) or molar heat capacity of a liquid determines how rapidly the temperature rises. Recall the following definition.

specific heat   -   the amount of heat required to raise the temperature of one gram of a substance one degree Celsius, *with no change in state.*

Example 13-2:   How much heat is released by 200. g of H_2O as it cools from 85.0°C to 40.0°C? The specific heat of H_2O is 4.18 J/g°C.

The molar heat capacity is the amount of heat required to raise the temperature of one mole of a substance one degree Celsius, *with no change in state*.

Example 13-3:   The molar heat capacity of ethyl alcohol, C_2H_5OH, is 113 J/mol°C. How much heat is required to raise the temperature of 125 grams of ethyl alcohol from 20.0°C to 30.0°C? 1 mol C_2H_5OH = 46.0 g.

Let us now examine the amounts of heat associated with changes of state.

heat of vaporization   -   the amount of heat that must be absorbed to convert one gram of a liquid *at its boiling point* to a gas *with no change in temperature*, usually J/g.

186

heat of condensation   -   the reverse of heat of vaporization; the amount of heat that   must be *removed* to liquefy one gram of a gas at its condensation (boiling) point *with no change in temperature.*

molar heat of vaporization, ΔH_{vap},   -   the amount of heat that must be absorbed to convert one mole of a liquid at *its boiling point* to a gas with no change in temperature.

molar heat of condensation   -   the reverse of the molar heat of vaporization

Numerical values are tabulated in Table 13-5 and Appendix E.

Example 13-4:   How many joules of energy must be absorbed by 500. g of H_2O at 50.0°C to convert it to steam at 120.°C? The molar heat of vaporization of water is 40.7 kJ/mol and the molar heat capacities of liquid water and steam are 75.3 J/mol°C and 36.4 J/mol°C, respectively.

Example 13-5:   If 45.0 g of steam at 140.°C is slowly bubbled into 450. g of water at 50.0°C in an insulated container, can all the steam be condensed?

We have described some properties of liquids and shown how they depend on *inter*molecular forces of attraction. The general effects of these attractions on physical properties of liquids are summarized in Table 13-6. Study it carefully.

Boiling Points Versus Intermolecular Forces

The boiling point of a liquid gives some indication of the strength of the attractive forces between individual particles in liquids, as well as their molecular weights. In general, the stronger the intermolecular forces, the higher the boiling point of the substance. *Hydrogen bonded* liquids have abnormally high boiling points.

Example 13-6:   Arrange the following substances in order of increasing boiling points.

C_2H_6, NH_3, Ar, NaCl, AsH_3

_____ < _____ < _____ < _____ < _____

THE SOLID STATE

13-10   Melting Point

The normal melting point of a solid (freezing point of its liquid) is the temperature at which liquid and solid coexist at equilibrium under a pressure of one atmosphere. Melting points of solids generally increase as the forces of attraction between the individual particles increase. Variations in melting and boiling points of substances are usually parallel.

13-11   Heat Transfer Involving Solids

heat of fusion, ΔH fusion, -   the amount of heat required to melt one gram of a solid at its melting point with no change in temperature. For example, the heat of fusion of ice is 334 joules per gram. Note that this is very large compared to the specific heat of water but not as large as the heat of vaporization of water (2.26×10^3 J/g). The molar heat of fusion, ΔH_f, is the amount of heat required to melt one mole of a substance at its melting point. See Figure 13-15, Table 13-7, and Appendix E.

heat of crystallization -   the amount of heat liberated by the crystallization of one
(solidification)   gram of liquid at its freezing point; it is equal to the amount of heat required to melt one gram of the solid at its melting point.

Let us summarize the heats of transformation for water. See Figure 13-15.

Example 13-7: Calculate the amount of heat required to convert 150.0 g of ice at -10.0°C to water at 40.0°C.

13-12 Sublimation and the Vapor Pressure of Solids

Sublimation is the conversion of a solid directly to vapor. "Dry Ice" is solid carbon dioxide. The white vapors we see around dry ice are due to water condensing in the very cold gaseous CO_2 near the solid. At a pressure of one atmosphere this conversion by-passes the liquid state. The reverse process is called *deposition*. See Figure 13-16.

$$\text{solid} \underset{\text{deposition}}{\overset{\text{sublimation}}{\rightleftharpoons}} \text{gas}$$

13-13 Phase Diagrams (P *vs*. T)

A phase diagram shows the equilibrium pressure-temperature relationships among the different phases of a given substance. Let us describe the phase diagram for _____ and _____. See Figure 13-17.

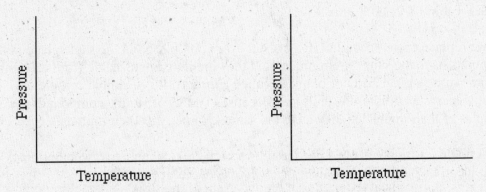

triple point   -   the point at which three phases of a substance – most commonly solid, liquid, and gas - can coexist in equilibrium

critical temperature -   the temperature above which a gas cannot be liquefied, i.e., the temperature above which the liquid and gas do *not* exist as distinct phases.

supercritical fluid - a substance above its critical temperature.

critical pressure - the pressure required to liquefy a gas at its critical temperature.

critical point - the combination of critical temperature and critical pressure.

13-14 Amorphous Solids and Crystalline Solids

Amorphous solids have no well-defined ordered structure. Examples include

X-ray diffraction can be used to determine the positions of atoms, ions, and molecules in solids. See the box on p. 472 and 473 and Figures 13-19 and 13-20.

Most solids are *crystalline*. They are characterized by definite repeating arrangements of particles.

13-15 Structures of Crystals

Crystals are polyhedra that consist of regularly repeating arrays of atoms, ions, or molecules. They are similar, in three dimensions, to a wallpaper pattern. The smallest unit of volume of a crystal that shows all the geometric characteristics of its pattern is called a *unit cell* (Figure 13-21). A three dimensional analog to a unit cell is a brick. By repeating a pattern of bricks, it is possible to make other structures; walls, houses, buildings; that have unique shapes and structures.

There are *seven* crystal systems and they are distinguished by the relative lengths of their axes (a, b, c) which are determined by the spacings between layers (d), and the angles between axes (α, β, γ). See Table 13-9 and Figure 13-23.

Suppose we replace each repeat unit in a crystal by a point placed in the same place in the unit. All such points have the same environment and they are indistinguishable from each other. The resulting three-dimensional array of points is called a *lattice*. It is a simple complete description of how a crystal structure is built up. Because crystals of simple compounds are repetitive multiples of their unit cells, they have the same symmetry as their unit cells.

Let us describe a few unit cells in some detail. In each case, all spheres (whether black or white) represent identical particles. Consider the *simple cubic* lattice. Study Figures 13-24, 13-25 and 13-26.

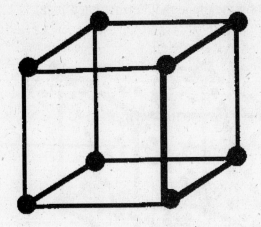

cubic, simple

The *cubic, body-centered* lattice differs from the simple cubic lattice in that an additional atom, ion, or molecule is in the center of the unit cell.

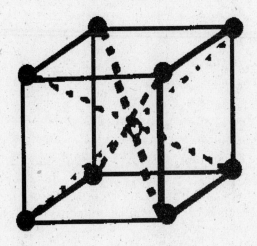

cubic, body-centered

The *cubic, face-centered lattice* differs from the simple cubic lattice in that an additional atom, ion, or molecule is bisected by the plane that defines each face.

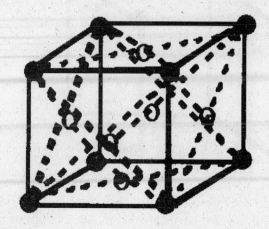

Cubic, face-centered

Example 13-8:   A group 4A element with a density of 11.35 g/cm^3 crystallizes in a face-centered cubic lattice whose unit cell edge length is 4.95 Å. Calculate the element's atomic weight. What is the atomic radius of this element?

Two important terms used in the descriptions of crystal structures are

192

isomorphous   -

polymorphous -

13-16   Bonding in Solids

One useful classification of solids is based on the kinds of particles and kinds of bonding. Table 13-10 lists some characteristics of four classes of solids.

1. **Metallic solids**, as the name implies, are metals. Bonding in metals is referred to as metallic bonding. Each valence electron in a piece of metal (no matter how large or small) is thought of as belonging to the "crystal". Thus, the picture of metals is of positively charged nuclei imbedded in a sea of electrons. Lattice points are *occupied by positively charged ions*. The excellent electrical conductivity of metals is taken as evidence for the mobility of electrons in metals. See Figure 13-36.

2. **Ionic solids** consist of positive and negative ions arranged in a definite crystal structure. The electrostatic attractions between oppositely charged ions are strong. Consequently, ionic solids are relatively high melting, hard, brittle solids. They are insulators. See Figures 13-28 and 13-29. Some examples are

3. **Molecular solids** consist of discrete molecules that occupy positions in unit cells. The attractive forces between the individual molecules are relatively weak. Molecular solids are usually soft with low melting points. They are volatile (as solids go) and are insulators. See Figure 13-31. Examples include

4. **Covalent solids** are also referred to as "network solids" or "giant molecules" because individual atoms are covalently bonded to several other atoms. Covalent solids should not be confused with covalent compounds--*simple covalent compounds usually form molecular solids*. Because each atom in a covalent solid is bonded to several other atoms, covalent solids are very hard, very high melting substances. Most are nonconductors. See Figure 13-32. Typical examples include

Allotropic modifications (allotropes) are different forms of the same element in the same physical state. Diamond and graphite (different forms of carbon) are common examples. See Figure 13-32.

The melting point of a solid gives some information about the strengths of the attractive forces among the individual particles that make up the solid. The stronger the attractive forces in the solid, the higher the melting point will be.

193

Molecular solids have low melting points (most < 300°C) because the attractive forces among molecules are rather weak.

Covalent solids melt at high temperatures (most > 1550°C) because the attractive forces among the individual particles are very strong.

Ionic solids melt at fairly high temperatures (most > 500°C) because the attractive forces among oppositely charged ions are much stronger than in molecular solids, but weaker than in covalent solids. Attractive forces increase as charges on ions increase and as their radii decrease.

Melting points of *metallic solids* (metals) vary widely because there are wide variations in the strengths of metallic bonding. Most metals have fairly high melting points, but mercury is a liquid at room temperature.

13-17   Band Theory of Metals

Many of the properties of metals such as metallic luster and high thermal and electrical conductivity can be explained in terms of the band theory of metallic bonding. Most metals crystallize in close-packed lattices (see Figures 13-26 and 13-27), which suggests that the attractive forces among atoms are strong. The representative metals, as well as most *d*-transition metals, contain too few valence electrons to participate in bonding with all their nearest neighbors. Therefore, we postulate a different kind of bonding to explain the observed facts.

Consider potassium as an example. Suppose that two atomic orbitals such as the 4*s* orbitals on two potassium atoms interact to produce two molecular orbitals, one bonding and one antibonding (Chapter 9).

Consider one mole of potassium atoms. The interactions of 6.022×10^{23} atomic 4*s* orbitals produce 6.022×10^{23} molecular orbitals. This suggests the possibility of a *band* of molecular orbitals with different energies because atoms that are closer together interact more strongly than atoms that are widely separated. See Figure 13-33(a).

194

Additionally, the vacant $4p$ (as well as $3d$) orbitals on potassium atoms can interact to form a wide band of molecular orbitals. See Figure 13-33(b). Because the energies associated with the various kinds of possible molecular orbitals overlap, electrons are able to move from one band to another resulting in a *conduction band*. This explains the ability of metallic potassium to conduct electricity when an electric field is applied.

Consider one mole of calcium as another example. It forms molecular orbitals resulting from the overlap of filled $4s$ and vacant $4p$ atomic orbitals. See Figure 13-34.

By contrast, *insulators*, i.e., nonconductors, have large *forbidden zones* that separate the valence band and the conduction band. Most nonmetals are insulators, although graphite does conduct electric current. See Figure 13-35.

Semiconductors do not conduct electricity well at room temperature because the valence and conduction bands are separated by a forbidden zone that is too wide, although not so wide as for most nonmetals. By increasing the temperature somewhat, some electrons acquire enough energy to move through the forbidden zone. See Figure 13-35. Raising the temperature of a metal significantly decreases its electrical conductivity, due to increasing collisions among the electrons.

Synthesis Question

Maxwell House Coffee Company decaffeinates its coffee beans using an extractor that is 7.0 feet in diameter and 70.0 feet long. Supercritical carbon dioxide at a pressure of 300.0 atm and temperature of 100.0°C is passed through the stainless steel extractor. The extraction vessel contains 100,000 pounds of coffee beans soaked in water until they have a water content of 50%. This process removes 90% of the caffeine in a single pass of the beans through the extractor. Carbon dioxide that has passed over the coffee is then directed into a water column that washes the caffeine from the supercritical CO_2. How many moles of carbon dioxide are present in the extractor?

Group Question

How many CO_2 molecules are there in 1.0 cm^3 of the Maxwell House Coffee Company extractor? How many more CO_2 molecules are there in a cm^3 of the supercritical fluid in the Maxwell House extractor than in a mole of CO_2?

Chapter Fourteen

SOLUTIONS

In Chapter 6 a solution was defined as a homogeneous mixture of two or more substances. A solution consists of a solvent (dissolving medium) and one or more solutes (dissolved species).

THE DISSOLUTION PROCESS

14-1   Spontaneity of the Dissolution Process

Review the solubility guidelines in Section 4-2 part 5. We shall now describe the major factors that influence solubility. We focus on dissolution processes in which no irreversible reaction occurs.

Because liquid solutions are the most common, we shall describe solutions in which the solvent is a liquid. Two major factors must be considered in determining the ease of dissolution of solutes (when no irreversible reaction with the solvent occurs). They are

1.   Change in energy content, the heat of solution, $\Delta H_{solution}$

2.   Change in disorder, or randomness, of the system

Solutes nearly always become more disordered upon dissolution, so let us consider the main factors that determined the *heat of solution* $\Delta H_{solution}$. They are

1.   Solute - solute attractions

2.   Solvent - solvent attractions

3.   Solvent - solute attractions

Dissolution is favored when the first two factors are small and the third is large. See Figure 14-1.

If there is an input of energy, the process is called endothermic, and if there is a release of energy, the process is exothermic.

14-2   Dissolution of Solids in Liquids

Many solids that are appreciably soluble in water are ionic. The *crystal lattice energy* is a measure of the solute - solute interactions for a solid solute, (1) above. It is defined as the amount of energy *absorbed* (endothermic process) when a mole of formula units of a solid is separated into its constituent ions (molecules or atoms for nonionic solids) in the gas phase. See Figure 14-1.

Crystal lattice energy is a measure of attractive forces among the particles of a solid. It increases (and the above process becomes more endothermic) as the charges on ions increase and as ionic radii decrease, i.e., as *charge density* increases. Crystal lattice energies of ionic solids are usually much higher than those of molecular solids.

Energy is also required to overcome solvent-solvent interactions such as hydrogen bonding in water, (2) above.

However, energy is released in the solvation of solute particles, (3) above. This is a measure of solute-solvent interactions. When the solvent is water, solvation is called hydration. All ions are hydrated in aqueous solution. Consider the dissolution of solid calcium chloride ($CaCl_2$), an ionic compound, in H_2O. See Figure 14-2.

The molar energy of hydration is the amount of energy absorbed when one mole of formula units, in the form of gaseous ions, becomes hydrated. In general terms

Like crystal lattice energy, hydration energy increases with increasing charge density. Consider the following examples. See Table 14-1 for more.

Ion	Ionic radius, (Å)	Charge/Radius Ratio	Heat of Hydration (kJ/mol)
K+	1.33	0.75	_____
Ca²⁺	0.99	2.02	_____
Cu²⁺	0.72	2.78	_____
Al³⁺	0.50	6.00	_____

Hydration energies and crystal lattice energies are often nearly equal, and so they tend to cancel each other for *low-charge, ionic solids*. Most such solids dissolve in water in endothermic processes due to the solvent-solvent interactions that must be overcome. The process is still spontaneous (favorable) due to the great increase in the disorder of the solute particles as they dissolve. See Figure 14-1.

Crystal lattice energies increase much faster than hydration energies as charge densities increase. Dissolution processes are very endothermic for many ionic solids that contain highly-charged ions, e.g., Al_2O_3, $Cr(OH)_3$, and Sb_2S_3. The endothermicity overcomes the effect of increased disorder, and so most such compounds are insoluble in water.

Nonpolar molecular solids such as naphthalene, $C_{10}H_8$, do not interact appreciably with water or other polar solvents. Nonpolar solids are usually soluble in nonpolar solvents such as carbon tetrachloride, CCl_4, and benzene, C_6H_6. This is because there are only weak attractive forces between solute molecules and between solvent molecules. Their solvation energies are very low, so both (1) and (2) above are very small.

Polar solvents dissolve ionic and polar molecular solutes, and nonpolar solvents dissolve nonpolar molecular solutes ("like dissolves like").

14-3 Dissolution of Liquids in Liquids (Miscibility)

Liquids that are soluble in each other are said to be *miscible*. Most polar liquids are miscible with other polar liquids, but immiscible with nonpolar liquids.

Most nonpolar liquids are miscible with other nonpolar liquids, but immiscible with polar liquids.

Thus, we can see reasons for the generalization that *"like dissolves like"*. Many ionic and polar covalent solutes dissolve in polar solvents such as water. Many nonpolar solutes dissolve in nonpolar solvents such as benzene and carbon tetrachloride.

Because solute-solute interactions are small for most liquid solutes, liquids usually dissolve in other liquids in *exothermic processes*. See Figure 14-4.

14-4   Dissolution of Gases in Liquids

Polar gases are more soluble in water than nonpolar gases because many polar gases hydrogen bond with water. Some polar gases such as HBr and SO_2 react with water as they dissolve.

Some nonpolar gases are reasonably soluble in water because they react with water.

Because solute-solute interactions are virtually nonexistent for gaseous solutes, gases dissolve in liquids in *exothermic processes*.

14-5   Rates of Dissolution and Saturation

A finely divided solid dissolves more rapidly than large crystals of the same substance. This is because a finely divided solid has a much larger surface area exposed to the solvent. Consider a single crystal of sodium chloride (a perfect cube) whose edges are 1 cm.

A solid solute dissolves in a given solvent until the solution becomes *saturated*, i.e., equilibrium exists between dissolved and undissolved solute.

Supersaturated solutions contain higher than saturation concentrations of solute. Supersaturated solutions of some solutes can be prepared by saturating a solution at a high temperature. The solution is then allowed to cool slowly, without agitation. A supersaturated solution represents a metastable condition. Almost any disturbance results in crystallization of the excess solute. See Figure 14-5.

14-6   Effect of Temperature on Solubility

LeChatelier's Principle (as seen in Section 13-6) states that when a stress is applied to a system at equilibrium (such as a saturated solution) the system responds in a way that best relieves the stress. Exothermic dissolution processes release heat and endothermic processes absorb heat.

202

Exothermic

Endothermic

Many ionic solids dissolve by endothermic processes. Their solubilities usually increase as temperature increases. The endothermic dissolution process consumes heat and relieves the "stress" of raising the temperature. See Figure 14-6. Some ionic compounds are more soluble in cold water than in hot water.

Most liquids and gases dissolve in liquids in exothermic processes. Their solubilities usually decrease as temperature increases. The reverse process consumes heat and relieves the stress of raising the temperature.

14-7   Effect of Pressure on Solubility

Changes in pressure have no appreciable effect on solubilities of solids or liquids in liquids. However, increases in partial pressures of gases favor the solubilities of gases in liquids (Henry's Law). The dissolution process reduces the partial pressure of the gas and relieves the stress of increased pressure. See Figure 14-7. Symbolically, Henry's Law may be represented as

14-8   Molality and Mole Fraction

Concentrations of solutions were introduced in Chapter 3 in terms of (1) percent by mass of solute, and (2) molarity. You should review those discussions and calculations.

Physical properties of solutions are conveniently related to their concentrations expressed as *molality (m)* or as a *mole fraction (X)*.

Molality

Molality is defined as the *number of moles of solute per kilogram of solvent*.

Note that molality is the number of moles of solute in a specified amount (1 kg) of *solvent. For very dilute aqueous solutions*, molarity and molality are very nearly equal. They are *not* for other solvents and more concentrated aqueous solutions.

Example 14-1: Calculate the molarity and the molality of an aqueous solution that is 10.0% glucose, $C_6H_{12}O_6$. The density of the solution is 1.04 g/mL. 10.0% glucose solution has several medical uses. 1 mol $C_6H_{12}O_6$ = 180 g.

Example 14-2: Calculate the molality of a solution that contains 7.25 grams of benzoic acid, C_6H_5COOH, in 200. mL of benzene, C_6H_6. The density of benzene is 0.879 g/mL. 1 mol C_6H_5COOH = 122 g.

Mole Fraction

The *mole fraction*, X_A and X_B, of each substance in a solution containing substances A and B is defined as

Mole fractions are dimensionless quantities. In any solution, the sum of the mole fractions must equal one.

204

Example 14-3:   What are the mole fractions of glucose and water in 10.0% glucose solution (Example 14-1)?

COLLIGATIVE PROPERTIES OF SOLUTIONS

Colligative properties of solutions depend on the number, rather than the kind, of solute particles. Colligative properties are physical properties of solutions. Lowering of vapor pressure, freezing point depression, boiling point elevation, and osmotic pressures are examples of colligative properties.

14-9   Lowering of Vapor Pressure and Raoult's Law

When a *nonvolatile* solute is dissolved in a liquid, the vapor pressure of the liquid is always lowered. This is because there are fewer solvent molecules per unit area on the surface of the liquid. Vapor pressure depends upon the ease with which molecules can escape from the surface of a liquid. This situation is described by *Raoult's Law*: *The vapor pressure of a solvent in an ideal solution decreases as its mole fraction decreases.* See Figure 14-8.

$$P_{solvent} = (X_{solvent})(P^o_{solvent})$$

where: $P_{solvent}$ =

$P^o_{solvent}$ =

$X_{solvent}$ =

The lowering of vapor pressure, $\Delta P_{solvent}$, is defined as

Because $(X_{solvent} + X_{solute}) = 1$, we have

Solutions that obey this relationship exactly are called *ideal* solutions. See Figures 14-8, 14-9, 14-10, and 14-11.

14-10   Fractional Distillation

Simple distillation is not very efficient for a solution that contains a *volatile* solute. Fractional distillation is a technique used to separate components of a solution that contain two or more volatile components. Two volatile components have different boiling points, and therefore different vapor pressures at a given temperature. See Figures 14-11 and 14-12. A fractionating column has a very large internal surface area on which condensation occurs.

14-11   Boiling Point Elevation

A solution that contains a *nonvolatile* solute always has a higher boiling point than the pure solvent. This is because the *nonvolatile* solute lowers the vapor pressure of the solvent. So a higher temperature is required for the vapor pressure of the solution to become equal to atmospheric pressure.

In accord with Raoult's Law, the elevation of the boiling point caused by a nonvolatile solute is proportional to the number of moles of solute dissolved in a given mass of solvent. See Figure 14-14. The relationship is

ΔT_b =   the boiling point elevation of the solvent

m =   the molality of the solute

K_b =   the molal boiling point elevation constant for the solvent

K_b is different for different solvents.

All *one molal **aqueous*** solutions of *nonvolatile nonelectrolytes* boil at 100.512°C at one atmosphere pressure. The number 0.512°C/m is the molal boiling point elevation constant, K_b, for water. Similarly, all *one molal solutions* of nonvolatile nonelectrolytes in *benzene* boil at 82.63°C at one atmosphere pressure. This is 2.53°C above the normal boiling point of benzene (80.10°C). Thus, the molal boiling point elevation constant for benzene is 2.53°C/m. See Table 14-2.

Example 14-4:   What is the normal boiling point of a 2.50 m glucose, $C_6H_{12}O_6$, solution?

14-12 Freezing Point Depression

The freezing (melting) point of a substance is the temperature at which the liquid and solid phases coexist in equilibrium. In a solution, the solvent molecules are somewhat more separated from each other by the solute molecules. The freezing point of a solution is always lower than that of the pure solvent. See Figure 14-14.

All aqueous solutions freeze at temperatures below 0°C. All one molal **aqueous** solutions of nonelectrolytes (covalent compounds that do not ionize) freeze at -1.86°C. The number of 1.86°C/m is the *molal freezing point depression constant*, K_f, for water. Similarly, one molal solutions of *nonelectrolytes* in benzene freeze at 0.36°C, which is 5.12°C lower than the freezing point for pure benzene (5.48°C). The molal freezing point depression constant for benzene is 5.12°C/m. See Table 14-2 and below.

Boiling Points, Boiling Point Elevation Constants, Freezing Points, and Freezing Point Depression Constants for Some Common Solvents

Solvent	Boiling Point (760 torr) °C	K_b °C/m	Freezing Point °C	K_f °C/m
Water	100.00	0.512	0.00	1.86
Carbon tetrachloride	76.75	5.03	-22.99	-
Benzene	80.10	2.53	5.48	5.12
Camphor	207.42	5.61	178.40	40.0

The relationship between freezing point depression and molality is

$$\Delta T_f = \text{freezing point depression for the solvent}$$

$$m = \text{molality of the solute}$$

$$K_f = \text{freezing point depression constant for the solvent}$$

Example 14-5: Calculate the freezing point of a 2.50 m aqueous glucose solution.

Example 14-6   Calculate the freezing point of a solution that contains 8.50 g of benzoic acid
(C_6H_5COOH, MW = 122) in 75.0 grams of benzene, C_6H_6.

14-13 Determination of Molecular Weights by Freezing Point Depression or Boiling Point Elevation

Molal freezing point depression constants are known for many solvents (Table 14-2). Freezing point depression measurements are frequently used to determine molecular weights of new compounds. A solvent for which K_f is known is selected. A carefully weighed sample of the new compound is dissolved in a carefully weighed amount of the solvent. The freezing point depression of the resulting solution is determined as accurately as possible.

Example 14-7:   A 37.0 g sample of a new covalent compound, a nonelectrolyte, was dissolved in 200. g of water. The resulting solution froze at -5.58°C. What is the molecular weight of the compound?

14-14 Colligative Properties and Dissociation of Electrolytes

Aqueous solutions of electrolytes conduct electric current because they contain ions. Freezing point depressions for solutions of electrolytes are larger than those for solutions of nonelectrolytes of the same concentrations. This is because the number of solute particles is greater in solutions of electrolytes than in solutions of nonelectrolytes of equal molar concentrations.

An electrolyte never behaves as if it were exactly 100% ionized or dissociated in aqueous solutions because of association of ions, or ion-pairing. At any instant, a small (statistical) percent of cations and anions will have collided and "stuck together". Such "ion pairs", or "associated ions", behave as it they were un-ionized for a brief period. See Figure 14-15. The extent of ion association decreases as concentration of the electrolyte decreases because collisions between ions occur less frequently.

Because of ion association, the observed freezing point depressions for solutions of electrolytes are never as great as expected if we assume that electrolytes are completely ionized or dissociated. Recall that freezing point depression is a colligative property, i.e., it depends on the number and not the kind of solute particles.

The following tabulation lists observed freezing point depressions and "expected" freezing point depressions for some solutions. In more dilute solutions of electrolytes, observed freezing point depressions are closer to "expected" depressions.

Observed Freezing Point Depressions Compared to "Expected" Depressions for Some Aqueous Solutions (assuming complete dissociation)

Solute		Concentrations of Solutions	
		0.00100 m	0.100 m
sucrose (sugar)	Measured	0.00186°C	0.186°C
nonelectrolyte	Expected	0.00186	0.186
NaCl	Measured	0.00366	0.348
2 ions/formula unit	Expected	0.00372	0.372
K_2SO_4	Measured	0.00528	0.432
3 ions/formula unit	Expected	0.00558	0.558
$K_3[Fe(CN)_6]$	Measured	0.00710	0.530
4 ions/formula unit	Expected	0.00744	0.744

The von't Hoff factor, i, is a measure of the extent of ionization or dissociation of an electrolyte in an aqueous solution.

The ideal, or limiting, value of i is 2 for 1:1 electrolytes such as NaCl and $MgSO_4$. For 2:1 (K_2SO_4) and 1:2 ($CaCl_2$) electrolytes, the ideal, or limiting, value of i is 3.

Example 14-8:   The freezing point of 0.0100 m NaCl solution is -0.0360°C. Calculate the van't Hoff factor and apparent percent dissociation of NaCl in this aqueous solution.

Example 14-9:   A 0.0500 m acetic acid solution freezes at -0.0948°C. Calculate the percent ionization of CH_3COOH in this solution.

14-15   Osmotic Pressure

When a dilute and a concentrated solution are separated by a semipermeable membrane, the solvent passes through the membrane *from* the solution of lower solute concentration *into* the solution of higher solute concentration. This spontaneous process is called *osmosis*. Cellophane is a semipermeable membrane. Many plant and animal membranes are semipermeable.

The solvent passes through the membrane in both directions, but more rapidly *from* the dilute solution *into* the concentrated solution than in the reverse direction. The initial difference in the two rates is directly proportional to the differences in concentrations of the two solutions (Figure 14-16). The column of liquid rises until the hydrostatic pressure due to the weight of solvent (water) in the column is sufficient to force solvent molecules back through the membrane at the same rate they enter from the dilute (or pure H_2O) side. The pressure exerted by the column of solvent is the *osmotic pressure* of the solution.

One mole of a nonelectrolyte in one kilogram of water, a 1.0 m solution, produces an osmotic pressure of approximately 22.4 atmospheres, nearly 330 p.s.i.! This pressure will support a column of water more than 700 feet high at 0°C! Osmotic pressures are very large, except for very dilute solutions.

$$\pi = \text{osmotic pressure in atmospheres}$$

$$M = \text{molarity of the solution}$$

$$R = 0.0821 \, \frac{L \bullet atom}{mol \bullet K}$$

In *very dilute aqueous solutions*, molality and molarity are very nearly equal.

(*For dilute aqueous solutions* **only**)

Osmotic pressure measurements are used to estimate molecular weights of complex materials with very high molecular weights, such as biological and other polymeric materials.

Example 14-10:   A 1.00 gram sample of a biological material was dissolved in enough water to give 100. mL of solution. The osmotic pressure of the solution was 2.80 torr at 25°C. Calculate the molarity and the approximate molecular weight of the material.

One important application of osmosis is the purification of water by reverse osmosis. Salt water is partially purified (some salt is removed) by this process.

COLLOIDS

Colloids (colloidal suspensions, or colloidal dispersions) represent mixtures intermediate between true solutions and mixtures that are classified as suspensions. The dispersed particles in colloids are sufficiently small (10-10,000 Å) that no settling, or any other kind of separation, occurs when a colloid is allowed to stand. See Table 14-4.

14-16   The Tyndall Effect

The Tyndall effect is a simple test for the "colloidal state". It is based on the scattering of light by colloidal particles.

There are many kinds of colloids. Table 14-4 contains a list and some examples.

14-17   The Adsorption Phenomenon

Because colloidal particles are so finely divided, they have very large surface areas in comparison to their volumes. Much of the chemistry of colloids is related to surface phenomena. The atoms on the surface of a colloidal particle are bonded to other atoms on and below the surface. They tend to interact strongly with other substances that come into contact with their surfaces. Consider the reaction of concentrated iron(III) chloride solution with hot water.

14-18   Hydrophilic and Hydrophobic Colloids

Colloids are classified as hydrophilic or hydrophobic.

Hydrophilic (water-loving) Colloids

Proteins are common examples.

Hydrophobic (water-hating) Colloids

An *emulsifying agent* is necessary to stabilize hydrophobic colloids in water.

Soaps and detergents are familiar examples of *emulsifying agents*.

soaps - sodium salts of long chain (fatty) organic acids (See page 536).

hard water - contains Fe^{3+}, Ca^{2+} and/or Mg^{2+} ions, all of which form "scum" with soap

synthetic detergents   -   soap-like emulsifying agents that usually contain sulfonate, $-SO_3^-$, or sulfate, $-OSO_3^-$, groups instead of $-COO^-$ groups

Synthetic detergents do not form "scum" because their Fe^{3+}, Ca^{2+}, and Mg^{2+} salts are soluble in water.

Synthesis Question

The world's record for altitude in flying gliders was 60,000 feet for many years. It was set by a pilot in Texas who flew into an updraft in front of an approaching storm. The pilot had to fly out of the updraft and head home not because he was out of air, there was still plenty in the bottle of compressed air on board, but because he was not wearing a pressurized suit. What would have happened to this pilot's blood if he had continued to fly higher?

Group Question

Medicines that are injected into humans, intravenous fluids and/or shots, must be at the same concentration as the existing chemical in our blood. For example, if the medicine contains potassium ions, they must be at the same concentration as the potassium ions in our blood. Such solutions are called *isotonic*. Why must medicines be formulated in this fashion?

Chapter Fifteen

CHEMICAL THERMODYNAMICS

Thermodynamics is the study of the changes and the transfers of energy that accompany chemical and physical processes. Thermodynamics addresses three fundamental questions.

1.   Will two (or more) substances react when they are mixed under specified conditions?

2.   If they do react, what energy changes and transfers are associated with their reaction?

3.   If a reaction occurs, to what extent does it occur?

We shall use thermodynamic data to determine *if* a reaction can occur under specified conditions. A reaction that can occur under the specified conditions is said to be *spontaneous*. Spontaneous reactions (in the thermodynamic sense) do not necessarily occur at an observable rate. *Nonspontaneous* reactions do not occur to a significant extent under the specified conditions.

HEAT CHANGES AND THERMOCHEMISTRY

15-1   The First Law of Thermodynamics

Energy is the capacity to do work or transfer heat.

Most, but not all, spontaneous reactions release energy in the form of heat, and are called *exothermic* reactions. Combustion reactions are familiar exothermic reactions. They are rapid oxidation-reduction reactions, accompanied by the liberation of heat and light. The combustion of propane and n-butane, common fossil fuels, are examples.

We have indicated specific amounts of heat as products of these reactions. These amounts of heat have been determined experimentally. The chemical potential energies of the products of *exothermic* reactions are lower than the potential energies of the *reactants*. See Fig. 15-1.

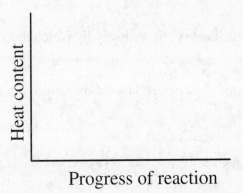

Two basic ideas are important in thermodynamics: (1) systems tend toward a state of *minimum potential energy*, and (2) systems tend toward a state of *maximum disorder*.

The *First Law of Thermodynamics* is also known as the Law of Conservation of Energy: *the total amount of energy in the universe is constant*. Earlier we pointed out the equivalence of matter and energy. The word "energy" in the first law includes the energy equivalent of all matter in the universe.

15-2 Some Thermodynamics Terms

System	the substances involved in the chemical and physical changes under investigation
Surroundings	the rest of the universe
Universe	the system and it surroundings
Thermodynamic state of a system	is defined by a set of conditions that completely specifies all the properties of the system. These commonly include the temperature, pressure, composition (identity and number of moles of each component), and the physical state (solid, liquid, or gas) of each part of the system. When the thermodynamic state of a system has been specified, all other properties - both chemical and physical - are fixed.

State functions   the properties of a system, e.g., pressure, energy, and temperature, whose values depend *only* on the state of the system. State functions are represented by capital letters.

State functions are independent of the pathway by which the system came to be in a given state.

Any property of a system that depends only on the value of its state functions is also a state function.

15-3   Enthalpy Changes

Most chemical and physical changes occur at constant pressure (usually at atmospheric pressure). Solution reactions in open beakers are examples of such reactions.

The **enthalpy change, ΔH**, of a process is defined as the quantity of heat transferred into or out of a system as it undergoes a chemical or physical change at constant pressure, q_p.

The enthalpy change of a reaction is casually referred to as the *heat change* or *heat of reaction*. The enthalpy change is equal to the enthalpy or "heat content" H, of the products minus the enthalpy of the reactants of the reaction.

For thermodynamic quantities, the *convention* is

$$\Delta(\qquad) = (\qquad)_{final} - (\qquad)_{initial}$$

where the parenthesis refer to the thermodynamic quantity of interest.

$$\Delta H = H_{final} - H_{initial} \quad \text{or} \quad \Delta H = H_{products} - H_{reactants}$$

$$\text{or} \quad \Delta H = H_{substances\ produced} - H_{substances\ consumed}$$

15-4   Calorimetry

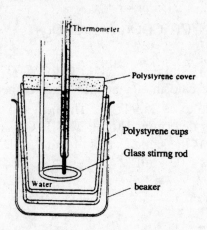

Thermometer

Polystyrene cover

Polystyrene cups

Glass stirrng rod

Water

beaker

A "coffee-cup" calorimeter is a device with which we can measure the amount of heat involved in a reaction at constant pressure, q_p, usually in aqueous solutions. Reactions are chosen so that there are no gaseous reactants or products. Therefore, all reactants and products remain in the container throughout the experiment. For an exothermic reaction, the amount of heat evolved by the reaction can be calculated from the amount of heat released which causes the temperature of the system to rise. The heat may be considered as divided in two parts.

$$\left\{ \begin{array}{c} \text{amount of heat} \\ \text{released by reaction} \end{array} \right\} = \left\{ \begin{array}{c} \text{amount of heat} \\ \text{gained by calorimeter} \end{array} \right\} + \left\{ \begin{array}{c} \text{amount of heat} \\ \text{gained by solution} \end{array} \right\}$$

The heat capacity of a calorimeter, or what is sometimes called the *calorimeter constant* is determined by measuring the rise in temperature of the calorimeter when a known amount of heat is added to the calorimeter. It depends on the construction materials of the calorimeter and its mass.

Example 15-1:   When 3.425 kJ of heat is added to a calorimeter containing 50.00 g of water the temperature rises from 24.00°C to 36.54°C. Calculate the heat capacity of the calorimeter in J/°C. The specific heat of water is 4.184 J/g °C.

1st: Find the temperature change.

2nd: Find the amount of heat absorbed by the water in going from 24.00°C to 36.54°C.

3rd: Find the amount of heat absorbed by the calorimeter.

4th: Find the heat capacity of the calorimeter
[=(amount of heat absorbed by calorimeter)/(temperature change)]

Example 15-2:   A coffee-cup calorimeter is used to determine the heat of reaction for the acid-base neutralization

$$CH_3COOH(aq) + NaOH(aq) \rightarrow NaCH_3COO(aq) + H_2O(\ell)$$

When we add 25.00 mL of 0.500 M NaOH at 23.000°C to 25.00 mL of 0.600 M CH_3COOH already in the calorimeter at the same temperature, the resulting temperature is observed to be 25.947°C. The heat capacity of the calorimeter has previously been determined to be 27.8 J/°C. Assume that the specific heat of the mixture is the same as that of water, 4.18 J/g°C and that the density of the mixture is 1.02 g/mL.

(1)   Calculate the amount of heat given off in the reaction.

(2)   Determine the number of moles of reactants consumed.

(3)   Calculate ΔH based on the limiting reactant calculation.

15-5   Thermochemical Equations

A balanced chemical equation, including its value of ΔH, is called a **thermochemical equation**. The equation

$$\underset{\text{1 mol}}{C_5H_{12}(\ell)} + \underset{\text{8 mol}}{8\,O_2(g)} \rightarrow \underset{\text{5 mol}}{5\,CO_2(g)} + \underset{\text{6 mol}}{6H_2O(\ell)} + 3523\text{ kJ}$$

is the thermochemical equation that describes the combustion (burning) of one mole of liquid pentane at a particular temperature and pressure. The coefficients in an equation such as this must be interpreted as *numbers of moles*. Thus, the equation says that 35.23 kJ of heat is released when one mole of $C_5H_{12}(\ell)$ reacts with eight moles of 8 $O_2(g)$ to give five moles of $CO_2(g)$ and six moles of $H_2O(\ell)$. This amount of reaction is often referred to as **one mole of reaction,** which is abbreviated as "mole rxn." This interpretation allows the writing of various unit factors.

$$\frac{1 \text{ mol } C_5H_{12}(\ell)}{1 \text{ mol rxn}}, \quad \frac{8 \text{ mole } O_2(g)}{1 \text{ mol rxn}}, \text{ and so on}$$

Notice that the introduction of a mole of reaction removes the necessity of specifying which reactant or product the ΔH_{rxn} is associated with. This amount of heat energy is the same whether we consider one mole of C_5H_{12} or eight moles of O_2 or five moles of CO_2, etc. All that is required is a balanced chemical reaction to specify the amount of heat associated with one mole of that reaction.

This thermochemical equation can also be written as

$$C_5H_{12}(\ell) + 8 O_2(g) \rightarrow 5 CO_2(g) + 6 H_2O(\ell) \qquad \Delta H = -3523 \text{ kJ/mol rxn}$$

The negative sign indicates that this is an *exothermic* reaction (it gives off heat).
The convention is to interpret the ΔH as the enthalpy for the reaction as written: that is, as (enthalpy change)/(mole of reaction), where the denominator means "for the number of moles of each substance shown in the balanced equation."

One of the keys to understanding thermochemical equation's is this mole of reaction concept. The following unit factors are important uses of this concept. These unit factors are

$$\frac{3523 \text{ kJ given off}}{\text{mol of reaction}} = \frac{3523 \text{ kJ given off}}{\text{mol } C_5H_{12}(\ell) \text{ consumed}} = \frac{3523 \text{ kJ given off}}{8 \text{ mol } O_2(g) \text{ consumed}}$$

$$= \frac{3523 \text{ kJ given off}}{5 \text{ mol } CO_2(g) \text{ formed}} = \frac{3523 \text{ kJ given off}}{6 \text{ H}_2O(\ell) \text{ formed}}$$

The reverse reaction would require the absorption of 3523 kJ under the same conditions; i.e., it is *endothermic* with $\Delta H = +3523$ kJ.

$$5CO_2(g) + 6 H_2O(\ell) \rightarrow C_5H_{12}(\ell) + 8 O_2(g) \qquad \Delta H = +3523 \text{ kJ/mol rxn}$$

It is important to remember the following conventions about thermochemical equations.

1. The coefficients in a balanced thermochemical equation refer to the numbers of *moles* of reactants and products involved. In the thermodynamic interpretation of equations we *never* interpret the coefficients as *numbers of molecules*. Thus, it is acceptable to write coefficients as fractions rather than as integers, when necessary.

2. The numerical value of ΔH (or any other thermodynamic quantity) refers to the *number of moles* of substances specified by the equation. This amount of change of substances is called *one mole of reaction*, so we can express ΔH in units of energy/mol rxn. For brevity, the units of ΔH are sometimes written kJ/mol or even just kJ. No matter what units are used, be sure that you interpret the thermodynamic change *per mole of reaction for the balanced chemical equation to which it refers*. If a different amount of material is involved in the reaction, then the ΔH (or other change) must be scaled accordingly.

3. The physical states of all species are important and must be specified.

4. The value of ΔH usually does not change significantly with moderate changes in temperature.

Example 15-3:   Write the thermochemical equation for the reaction in Example 15-2.

15-6   Standard States and Standard Enthalpy Changes

By international agreement thermodynamic data are tabulated for thermochemical standard state conditions, 298.15 K, and one atmosphere pressure. By convention ΔH^0_{298} refers to enthalpy changes measured at 1.00 atmosphere and 25°C (298.15 K). The conventions for thermochemical standard states are:

1. For a pure substance in the liquid or solid phase, the standard state is the pure liquid or solid.

2. For a gas, the standard state is the gas at one atmosphere pressure; in a mixture of gases, its partial pressure must be one atmosphere.

3. For a substance in solution, the standard state refers to *one-molar* concentration.

Examples of a few substances in their standard states and their standard molar enthalpies of formation include

$C_2H_5OH(\ell)$	-277.7 kJ/mol	$H_2O(\ell)$	-285.8
$CO_2(g)$	-393.51	$NaCl(s)$	-411.0

15-7 Standard Molar Enthalpies of Formation, ΔH_f^0

The *standard molar enthalpy (heat) of formation*, ΔH_f^0, is a change in enthalpy (change in heat content) for the reaction in which *one mole* of a substance in a specified state is formed from its elements in its standard states. For example

$$2 \, C(graphite) + 3 \, H_2(g) + {}^1\!/_2 \, O_2(g) \rightarrow C_2H_5OH(\ell) \qquad \Delta H_f^0 = -277.7 \text{ kJ/mol}$$

Standard molar enthalpies of formation for many compounds have been determined. A few are tabulated in Table 15-1. Appendix K contains many more.

The standard molar enthalpy of formation for an *element* in its most stable form at 25°C and one atmosphere of pressure is zero. Recall that the ΔH_f^0 refers to the amount of energy required to form one mole of a substance from its elements. Does it not seem logical that it should take zero energy to make an element from itself? Refer to Table 15-1 and to Appendix K in the text.

Example 15-4:   The standard molar enthalpy of formation for phosphoric acid is -1281 kJ/mol. Write the equation for the reaction for which $\Delta H_{rxn} = -1281$ kJ.

Example 15-5:   Calculate the enthalpy change for the reaction of one mole of $H_2(g)$ with one mole of $F_2(g)$ to form two moles of HF(g) at 25°C and one atmosphere.

$$H_2(g) \ + \ F_2(g) \ \rightarrow \ 2 \, HF \, (g)$$

223

Example 15-6:   Calculate the enthalpy change for the reaction in which 15.0 g of aluminum reacts with excess oxygen to form Al_2O_3 at 25°C and one atmosphere.

$$4 \, Al(s) \; + \; 3 \, O_2(g) \; \rightarrow \; 2 \, Al_2O_3(s)$$

15-8   Hess's Law

Hess' Law of heat summation states that the enthalpy change for a reaction is the same whether it occurs by one step or by any (hypothetical) series of steps (Figure 15-4). This is true because enthalpy is a state function.

As a proof of this idea, let us show that the $\Delta H°$ for a reaction can be determined from two other reactions. $\Delta H°$ is known to be -560 kJ for the reaction of four moles of FeO(s) with one mole of oxygen to form two moles of $Fe_2O_3(s)$.

		$\underline{\Delta H° \text{ (kJ)}}$
[1]	$4 \, FeO(s) + O_2(g) \; \rightarrow \; 2 \, Fe_2O_3(s)$	-560

$\Delta H°$ is also known for the two following reactions.

[2]	$2 \, Fe(s) + O_2(g) \rightarrow 2 \, FeO(s)$	-544
[3]	$4 \, Fe(s) + 3 \, O_2(g) \rightarrow 2 \, Fe_2O_3(s)$	-1648

If we had not known $\Delta H°$ for reaction [1] we could have calculated it by doubling and reversing equation [2] and then adding it to equation [3]. We do the same to the $\Delta H°$ values. (To show that we are doubling reation [2] we write 2[2]. To show that reaction [2] has been reversed we write [-2].)

		$\underline{\Delta H°}$	
$2 \times [-2]$			kJ
[3]			kJ
[1]		$\Delta H° =$	kJ

224

Example 15-7:   Given the following equations and $\Delta H°$ values,

$$\Delta H° \text{ (kJ)}$$

[1] $2 N_2(g) + O_2(g) \rightarrow 2 N_2O(g)$ 164.1

[2] $N_2(g) + O_2(g) \rightarrow 2 NO(g)$ 180.5

[3] $N_2(g) + 2 O_2(g) \rightarrow 2 NO_2(g)$ 66.4

calculate $\Delta H°$ for the reaction below.

$$N_2O(g) + NO_2(g) \rightarrow 3 NO(g) \qquad \Delta H° = ?$$

We want 1 mole of N_2O as a reactant, so we take $1/2$ of the reverse of eqn [1].
We want 3 moles of NO as the product, so we take $3/2$ of eqn [2].
We want 1 mole of NO_2 as a reactant, so we take $1/2$ of the reverse of eqn [3].
We do the same operations on the corresponding $\Delta H°$ values.

	$\Delta H°$
	kJ
$1/2 \times [-1]$	kJ
$3/2 \times [2]$	kJ
$1/2 \times [-3]$	kJ
$\Delta H° =$	kJ

This reaction is endothermic. The reverse reaction is exothermic, i.e.,

$$3 NO(g) \rightarrow N_2O(g) + NO_2(g) \qquad \Delta H° = \underline{\qquad} kJ$$

A more useful statement of Hess' Law is: for any chemical reaction at standard state conditions, the standard enthalpy change is the sum of the standard molar enthalpies of formation of the products (each multiplied by its coefficient in the balanced chemical equation) minus the corresponding sum for the reactants.

Example 15-8:   Calculate ΔH^0_{rxn} for the following reaction from data in Appendix K.

$$C_3H_8(g) + 5 O_2(g) \rightarrow 3 CO_2(g) + 4 H_2O(\ell)$$

225

ΔH^0_{rxn} = _____, and so the reaction is_____.

We can also use Hess' Law to calculate ΔH^0_f for a substance participating in a reaction for which we know ΔH^0_{rxn}, *if* we also know ΔH^0_f for all the other substances in the reaction.

Example 15-9:   Given the following information, calculate ΔH^0_f for $H_2S(g)$.

$$\Delta H^0_{rxn} = -1124 \text{ kJ}$$

		$2 H_2S(g)$	+	$3 O_2(g)$	$\rightarrow$	$2 SO_2(g)$	+	$2 H_2O(\ell)$
ΔH^0_f (kJ/mol)		?		0		-296.8		-285.8

We apply Hess' Law.

Now we solve for $\Delta H^0_{f\,H_2S}$.

15-9   Bond Energies

The strength of a chemical bond is measured by the amount of energy required to break one mole of bonds in a gaseous covalent substance to form products in the gaseous state at constant temperature and pressure, and is called the *bond energy*.

This is the *average* value of the several bond breaking reaction energies.

Table 15-2 shows average single bond energies for several common bonds. Double and triple bond energies are higher than single bond energies involving the same elements. A few are given in Table 15-3.

Molecule	Bond Energy (kJ/mol)	Molecule	Bond Energy (kJ/mol)
F_2	_____	O_2 (double bond)	_____
Cl_2	_____	N_2 (triple bond)	_____
Br_2	_____		

Bond energies can be calculated from other ΔH^0_{rxn} values.   See Figure 15-5.

Example 15-10:   Calculate the bond energy for hydrogen fluoride, HF.

Example 15-11.   Calculate the **_average_** N-H bond energy in ammonia, NH_3.

$\Delta H°$ values for reactions may also be related to the bond energies of all species involved in gas phase reactions by one form of Hess' Law. See Figure 15-5.

Example 15-12:   Use the bond energies listed in Tables 15-2 and 15-3 to estimate the heat of reaction at 25°C for the reaction below.

$$CH_4(g)   +   2 O_2(g)   \rightarrow   CO_2(g)   +   2 H_2O(g)$$

15-10   Changes in Internal Energy, ΔE

The internal energy, E, of a specific amount of a substance represents *all* the energy contained within the substance.

The first law of thermodynamics tells us that if some heat energy, q, is added to a system of *internal energy*, E_1, and the system in turn does some work, w, on the surroundings, the new internal energy of the system, E_2, is

q is + if   _____

q is - if   _____

w is + if   _____

w is - if   _____

ΔE is _____ when energy is *released* by a system undergoing a chemical *or* physical change. Energy can be written as a *product* of the process.

ΔE is _____ when energy is *absorbed* by a system undergoing a chemical *or* physical change. Energy can be written as a *reactant* of the process.

Example 15-13:   If 1200 joules of heat are added to a system in energy state, E_1, and the system does 800 joules of work on the surroundings, what is the

     (a)   energy change for the system, ΔE_{sys}?

     (b)   energy change of the surroundings, ΔE_{surr}?

     (c)   energy of the system in the new state, E_2?

     In most chemical and physical changes, the only kind of work is pressure-volume work.

Pressure is force per unit area

Volume is distance cubed.

$P\Delta V$ is a work term, (i.e., the same units are used for energy and work.)

The ideal gas law is $PV = nRT$. For a change in the volume of ideal gases at *constant temperature and pressure*, due to a change in the number of moles of gas molecules present, Δn_{gas},

Δn_{gas} = (no. of moles of gaseous products) - (no. of moles of gaseous reactants)

Work is defined as a force acting through a specific distance.

Many chemical reactions occur in open containers at atmospheric pressure, *at constant temperature and pressure*, the work term for chemical reactions is $P(V_2 - V_1) = P\Delta V$. See Figures 15-6, 15-7 and 15-8. Three possibilities exist.

	When	Then	Example
(1)	$V_2 = V_1$	$P\Delta V = 0$	
		$\Delta n_{gas} = 0$	
(2)	$V_2 > V_1$	$P\Delta V > 0$	
		$\Delta n_{gas} > 0$	
(3)	$V_2 < V_1$	$P\Delta V < 0$	
		$\Delta n_{gas} < 0$	

Consider the following gas phase reaction at *constant pressure* at 200°C.

$$2\ NO(g) \quad + \quad O_2(g) \quad \rightarrow \quad 2\ NO_2(g)$$

229

Consider the following gas phase reaction at *constant pressure* at 1000°C.

$$PCl_5(g) \quad \rightarrow \quad PCl_3(g) \quad + \quad Cl_2(g)$$

15-11   Relationship Between ΔH and ΔE

Because many chemical reactions occur in open containers (and outside containers for that matter) at atmospheric pressure, expansion work at constant pressure is a very useful concept. Many reaction systems undergo significant changes in volume at constant pressure. Adding the $P\Delta V$ work term to the change in internal energy for a system gives *the total amount of heat energy a reaction can provide to it surroundings*, i.e.,

$q_p = \Delta E + P\Delta V = \Delta H$ (definition of ΔH and thus H)

where: H   =   _____

ΔH =   _____

ΔE =   _____

$P\Delta V$ =   _____

For reactions in which the volume change is very small or equal to zero,

Example 15-14:   In Section 15-5. we found that $\Delta H^0 = -3523$ kJ/mol for the combustion of n-pentane, n-C_5H_{12}. Combustion of one mole of n-pentane at constant pressure releases 3523 kJ of heat. What are the values of the work term and ΔE for this reaction?

$$C_5H_{12}(\ell) \quad + \quad 8\,O_2(g) \quad \rightarrow \quad 5\,CO_2(g) \quad + \quad 6\,H_2O(\ell)$$

SPONTANEITY OF PHYSICAL AND CHEMICAL CHANGES

The concept of spontaneity has a very specific interpretation (meaning) in thermodynamics. A spontaneous change is one that can happen without any continuing outside influence. A spontaneous change has a natural direction.

rusting of iron

melting of ice at room temperature

15-12   The Two Aspects of Spontaneity

Two factors affect the spontaneity of any physical or chemical change:

1. Spontaneity is favored when heat is released during the change (exothermic).

2. Spontaneity is favored when the change causes an increase in the dispersal of energy and matter.

Often one of these factors "overrides" the other in a chemical reaction or a physical change.

15-13   Dispersal of Energy and Matter

Dispersal of Energy

The dispersal of energy in a system results in the energy being spread over many particles rather than being concentrated in just a few. If energy can be dispersed over a larger number of particles, it will be.

The greater the number of molecules and the higher the total amount of energy in the system are less likely that the energy be concentrated in a few molecules. The energy in turn will be more dispersed.

Dispersal of Matter

The dispersal of matter, as in the expansion of gas, also often contributes to energy dispersal. When the gas is allowed to occupy a larger volume, its energy levels become closer together than it was previously when a smaller volume was occupied. This causes more ways for the gas to disperse its energy to arrive at the same total energy.
The molecules of one gas can transfer energy to another gas when the two gasses collide. This will disperse energy along with dispersing matter. Dispersal of matter is described as an increase in disorder of the system.

15-14   Entropy, S, and Entropy Change, ΔS

The thermodynamic state function entropy, S, is a measure of the *disorder* or *randomness* of a system.

The *Third Law of Thermodynamics* states that the entropy of a pure, perfect, crystalline substance at 0K is zero. See Figure 15-15.

Entropies can be measured quantitatively. Absolute standard molar entropy values (S^0_{298}) are tabulated in Appendix K.

An increase in entropy corresponds to an increase in disorder. This is the principal driving force for many physical processes and chemical reactions. The conventions are

When: ΔS   is   _____ disorder increases   (favors spontaneity)

ΔS   is   _____ disorder decreases   (disfavors spontaneity)

Entropy is one of the most basic concepts in science. Generally, if a single substance is compared, gases (highly disordered) have higher entropies than liquids (less disordered), and liquids have higher entropies than solids (well ordered).

A relationship similar to Hess' Law is applicable to entropy changes and is used to find the standard entropy change, ΔS°.

Example 15-15:   Calculate the entropy change for the following reaction at 25°C.

$$2\ NO_2(g) \quad \rightarrow \quad N_2O_4(g)$$

232

ΔS^0_{rxn} is negative because the system becomes more ordered. For the reverse reaction,

$$N_2O_4(g) \quad \rightarrow \quad 2\ NO_2(g)$$

ΔS^0_{rxn} = _____ because disorder increases as this reaction occurs.

Example 15-16:   Calculate ΔS^0_{rxn} for the reaction below.

$$3\ NO(g) \quad \rightarrow \quad N_2O(g) \quad + \quad NO_2(g)$$

Changes in entropy are usually quite small compared to changes in internal energy (ΔE) and changes in enthalpy (ΔH).

15-14 The Second Law of Thermodynamics

The Second Law of Thermodynamics states: *In spontaneous changes the universe tends toward a state of greater disorder (increasing entropy),* $\Delta S_{universe} > 0$.

If entropy increases in a system, during a process, spontaneity is favored, but not required. The entropy of the universe (not the system) increases during a spontaneous process.

15-16 Free Energy Change, ΔG, and Spontaneity

If heat is released in a chemical reaction (i.e., ΔH is *negative*), some of the heat may be converted into useful work. Some of it may be used to increase the order of the universe (if ΔS is *negative*). If the system becomes more disordered (i.e., ΔS is *positive*), then more useful energy becomes available than indicated by ΔH alone.

J. Willard Gibbs formulated the relationship between changes in enthalpy and entropy for a process in terms of another state function, the *Gibbs free energy change, ΔG.*

(constant T and P)

233

The amount by which the Gibbs free energy decreases, is the maximum useful energy obtainable in the form of work from a process at constant T and P. Gibbs free energy is a state function, as are changes in free energy. ΔG_f^0 values (standard molar free energies of formation) are listed in Appendix K.

A Gibbs free energy change, ΔG *not* $\Delta G°$, is a reliable indicator of spontaneity for a physical change and for a chemical reaction. However, ΔG tells us nothing about how *fast* a process occurs.

When ΔG is > 0 _____

When ΔG is = 0 _____

When ΔG is < 0 _____

Strictly speaking, $\Delta G°$ values are applicable to a very special case that we shall call *the standard reaction*. In *the standard reaction*, the numbers of moles of reactants shown in the balanced equation, all in their standard states, are converted into the number of moles of products shown in the balanced equation, also in their standard states. Remember that ΔG *not* $\Delta G°$, is the indicator of spontaneity!

We may apply a relationship similar to Hess' Law to changes in free energy.

Example 15-17:   Calculate ΔG_{rxn}^0 for the reaction in Example 15-9.

$$C_3H_8(g) \;+\; 5\,O_2(g) \;\rightarrow\; 3\,CO_2(g) \;+\; 4\,H_2O(\ell)$$

ΔG_{rxn}^0 is _____, so the reaction is _____ under standard state conditions. For the reverse reaction,

$$3\,CO_2(g) \;+\; 4\,H_2O(\ell) \;\rightarrow\; C_3H_8(g) \;+\; 5\,O_2(g), \quad \Delta G° = _____$$

ΔG_{rxn}^0 is _____, so this reaction is _____ under standard state conditions.

234

15-17 The Temperature Dependence of Spontaneity

The quantitative difference between ΔG_{rxn}^0 and ΔH_{rxn}^0 is very nearly proportional to the absolute temperature and is related to the randomness or disorder of a system. As we pointed out earlier, the general relationship among ΔG, ΔH, and ΔS is

The relationships among the signs of ΔH and ΔS (and therefore ΔG) and the possibilities for spontaneity follow. There are four possibilities. Study Figure 15-21 and Table 15-7.

	ΔH	ΔS	ΔG	Therefore
(1)	___	___	___	forward reaction spontaneous at all T's
(2)	___	___	___	forward reaction spontaneous at low T's
(3)	___	___	___	forward reaction spontaneous at high T's
(4)	___	___	___	forward reaction nonspontaneous at all T's

Example 15-18:   Calculate ΔS_{rxn}^0 for the following reaction. In Example 15-8, we found that $\Delta H_{rxn}^0 =$ _____, and in Example 15-17 we found that $\Delta G_{rxn}^0 =$ _____.

$$C_3H_8(g) \ + \ 5\,O_2(g) \ \rightarrow \ 3\,CO_2(g) \ + \ 4\,H_2O(\ell)$$

$\Delta S_{rxn}^0 =$ _____ which indicates that the disorder of the system_____.

For the reverse reaction,

$$3\,CO_2(g) \ + \ 4\,H_2O(\ell) \ \rightarrow \ C_3H_8(g) \ + \ 5\,O_2(g)$$

$\Delta S_{rxn}^0 =$ _____ which indicates that the disorder of the system _____.

Example 15-19:   Use thermodynamic data to estimate the normal boiling point of water.

Example 15-20:   What is the percent error in Example 15-19?

Synthesis Question

When it rains an inch of rain, that means that if we built a one inch high wall around a piece of ground that the rain would completely fill this enclosed space to the top of the wall. Rain is water that has been evaporated from lakes, oceans, and rivers and then precipitated back on the land. How much heat must the sun provide to evaporate enough water to rain 1.0 inch on 1.0 acre of land? 1 acre = 43,460 ft^2

Group Question

When Ernest Rutherford, introduced in Chapter 5, gave his first lecture to the Royal Society one of the attendees was Lord Kelvin. Rutherford announced at the meeting that he had determined that the earth was at least 1 billion years old, 1000 times older than Kelvin had previously determined for the earth's age. Then Rutherford looked at Kelvin and told him that his method of determining the earth's age based upon how long it would take the earth to cool from molten rock to its present cool, solid form was essentially correct. But there was a new, previously unknown source of heat that Kelvin had not included in his calculation and therein lay his error. Kelvin apparently grinned at Rutherford for the remainder of his lecture. What was this "new" source of heat that Rutherford knew about that had thrown Kelvin's calculation so far off?

Chapter Sixteen

CHEMICAL KINETICS

Chemical kinetics is the study of *rates* of chemical reactions, the factors that affect reaction rates, and the *mechanisms* by which they occur.

A *reaction rate* describes how fast a reactants are used up and products are formed.

Chemical kinetics is concerned with the rates of reactions, but the kinetics of a reaction tell us nothing about the extent to which it can occur.

In chemical thermodynamics (Chapter 15), we studied the energy changes associated with chemical reactions. Thermodynamics tells us whether a reaction can occur under specified conditions. The extent to which a reaction can occur is addressed by thermodynamics, and not by kinetics. Some reactions that can occur under specified conditions occur very slowly. For all practical purposes, we say these reactions do not occur under the specified conditions.

16-1   The Rate of a Reaction

Consider the hypothetical reaction,

$$aA(g) \;+\; bB(g) \;\rightarrow\; cC(g) \;+\; dD(g)$$

The amount of each substance present can be given by its concentration, which is usually expressed in molarity. Because reaction rates describe the rates at which reactants disappear or at which products appear, we may write

The rate of a simple one-step reaction is directly proportional to the concentration of the reacting substance. For a *one-step* reaction, we represent the rate of reaction,

$$A(g) \quad \rightarrow \quad B(g) \quad + \quad C(g)$$

The square brackets represent molar concentrations, i.e., moles/liter. The proportionality constant, k, is called the specific rate constant. The rate-law expression, $R = k[A]$, describes the rate at which the reaction occurs at a specified temperature. Doubling the initial concentration of A doubles the initial rate of this reaction. Halving the initial concentration of A halves the initial rate of this reaction.

FACTORS THAT AFFECT REACTION RATES

Several factors influence the rates of chemical reactions. They will be discussed individually. These factors are (1) nature of reactants, (2) concentrations of reactants, (3) temperature, and (4) the presence of a catalyst. Obviously, the rate at which a reaction occurs depends upon the substances that are mixed together and the conditions under which they are mixed.

16-2   Nature of the Reactants

The "nature of the reactants" is a broad category that includes a number of properties of reacting substances. To illustrate the point, consider the following.

1. Sodium reacts with water explosively at room temperature to liberate hydrogen and form sodium hydroxide.

2. Calcium reacts with water only slowly at room temperature to liberate hydrogen and form calcium hydroxide.

3. The reaction of magnesium with water at room temperature is so slow that the evolution of hydrogen is not perceptible to the human eye.

However, Mg reacts with steam rapidly to liberate H_2 and form magnesium oxide.

The different rates at which sodium, calcium, and magnesium react with water at room temperature are attributed to differences in the "nature of the reactants".

16-3   Concentrations of Reactants: The Rate-Law Expression

Rate-law expressions (sometimes called rate laws, rate equations, or rate expressions) must be determined experimentally. **They cannot be written down by inspecting balanced chemical equations!** This is because most reactions are *not* one-step reactions.

The *order* of a reaction may be expressed in terms of the order with respect to each reactant *or* in terms of the order of the overall reaction. The following examples illustrate the point. These rate-law expressions have been *determined experimentally*.

$2 N_2O_5(g) \rightarrow 4 NO_2(g) + O_2(g)$ $R = k[N_2O_5]$

_____order in N_2O_5 and _____ order overall

$(CH_3)_3CBr(aq) + OH^- (aq) \rightarrow (CH_3)_3COH(aq) + Br^-(aq)$ $R = k[(CH_3)_3CBr]$

_____order in $(CH_3)_3CBr$ and _____order in OH^-, and _____ order overall

$2 NO(g) + O_2(g) \rightarrow 2 NO_2(g)$ $R = k[NO]^2[O_2]$

_____order in NO, _____order in O_2, and _____ order overall

242

Consider the following simplified representation of the effect of different numbers of molecules in a given volume (the *concentration*) on the rate of a simple one-step chemical reaction, A(g) + B(g) → Products.

Thus, we see that increasing the concentrations of reactants increases the probability of collision and the probability that reaction will occur.

Earlier, we pointed out the fact that for one-step reactions, such as A→ B + C, doubling the concentration of A doubles the reaction rate. Consider the following *one step* reaction and its experimentally determined rate-law expression

$$2 \text{ A(g)} \rightarrow \text{B(g)} + \text{C(g)} \qquad\qquad \text{Rate} = k[A]^2$$

This rate-law expression tells us that doubling the initial concentration of A increases the reaction rate by a factor of *four*. Halving the initial concentration of A decreases the initial rate by a factor of *four*, i.e.,

Let us now examine some typical kinetic data from which reaction rate-law expressions may be determined.

Example 16-1: The following rate data were obtained at 25°C for the following reaction. What are the rate-law expression and the specific rate constant for this reaction?

$$2 \text{ A(g)} + \text{B(g)} \rightarrow 3 \text{ C(g)}$$

Experiment	Initial [A]	Initial [B]	Initial Rate of Formation of C (M/s or $M \cdot s^{-1}$)
1	0.10 M	0.10 M	2.0×10^{-4}
2	0.20 M	0.30 M	4.0×10^{-4}
3	0.10 M	0.20 M	2.0×10^{-4}

Example 16-2:   These data were obtained for the following reaction at 25°C. What are the
                 rate-law expression and the specific rate constant for the reaction?

$$2 \text{ A(g)} \quad + \quad \text{B (g)} \quad + \quad 2 \text{ C(g)} \quad \rightarrow \quad 3 \text{ D(g)} \quad + \quad 2 \text{ E(g)}$$

Experiment	Initial [A]	Initial [B]	Initial [C]	Initial Rate of Formation of D (M/s or $M \cdot s^{-1}$)
1	0.20 M	0.10 M	0.10 M	2.0×10^{-4} M $\cdot$ s^{-1}
2	0.20 M	0.30 M	0.20 M	6.0×10^{-4} M $\cdot$ s^{-1}
3	0.20 M	0.10 M	0.30 M	2.0×10^{-4} M $\cdot$ s^{-1}
4	0.60 M	0.30 M	0.40 M	1.8×10^{-3} M $\cdot$ s^{-1}

Example 16-3:   Consider a chemical reaction between compounds A and B that is first
                 order in A and first order in B and second order overall. From the
                 information given below, fill in the blanks.

Experiment	Initial Rate (M/s or $M \cdot s^{-1}$)	Initial [A]	Initial [B]
1	4.0×10^{-3}	0.20 M	0.050 M
2	1.6×10^{-2}	____ M	0.050 M
3	3.2×10^{-2}	0.40 M	_____ M

244

16-4 Concentration versus Time: The Integrated Rate Equation

The integrated rate equation relates time and concentration for a chemical reaction. The integrated rate equation and half-life are different for reactions of different order. An integrated rate equation can also be used to calculate the half-life, $t_{1/2}$, of a reactant.

First-Order Reactions

For a reaction that is *first order in reactant A*, and *first order overall* the following integrated rate equation, applies

$[A]_o$ = mol/L of A at t = 0 (or number of moles or number of grams)

$[A]$ = mol/L of A at time t (or number of moles or number of grams)

k = the specific rate constant

t = time elapsed since beginning of reaction

a = stoichiometric coefficient of A in balanced equation

We can replace $[A]_o$ and $[A]$ with *amounts* of A in either moles or grams as long as the same unit is used consistently.

We can solve the relationship for t

The half-life, $t_{1/2}$, of a reactant is the time required for half of it to be consumed, or the time at which $[A] = 1/2[A]_o$. At this time $t = t_{1/2}$ and the above expression becomes

Because $\dfrac{[A]_o}{(1/2)[A_o]} = 2,$

ln 2 = 0.693, so $t_{1/2}$ is

AN IMPORTANT NOTE: The equations in this section apply only to reactions in which a, the coefficient of A in the balanced overall equation, is 1. To apply any of these equations to other cases, we must multiply the value of the specific rate constant, k, by the coefficient of A in the balanced equation.

Example 16-4:   Cyclopropane, an anesthetic, decomposes to propene according to the following equation.   The reaction is first order in cyclopropane with k = 9.2 s^{-1} at 1000°C. Calculate the half-life of cyclopropane at 1000°C:

We see that the reaction is extremely rapid at 1000°C!

Example 16-5:   Refer to Example 16-4. How much of a 3.0 g sample of cyclopropane remains after 0.50 second?

246

Example 16-6:   The half-life for the following first order reaction is 688 hours at 1000°C. Calculate the specific rate constant, k, at 1000°C and the amount of a 3.0 gram sample of CS_2 that remains after 48 hours.

$$CS_2(g) \quad \rightarrow \quad CS(g) \quad + \quad S(g)$$

Second-Order Reactions

For reactions that are *second order* with respect to a particular reactant and *second order overall*, the integrated rate equation is

For $t_{1/2}$ $[A] = (1/2)[A]_0$, so

The common denominator is $[A]_0$.

Solving for $t_{1/2}$ gives

Thus we see that the half-life of a *second order* reaction depends upon the *initial* concentration (or amount) of A.

Example 16-7:   Acetaldehyde, CH_3CHO, undergoes gas phase thermal decomposition to methane and carbon monoxide.

$$CH_3CHO(g) \rightarrow CH_4(g) + CO(g)$$

The rate-law expression is Rate = $k[CH_3CHO]^2$, and k = 2.0×10^{-2} L/(mol•hr) at 527°C.

(a)   What is the half-life of CH_3CHO if 0.10 mole of it is injected into a 1.0-L vessel at 527°C?

(b)   How many moles of CH_3CHO remain after 200 hours?

(c)   What percent of the CH_3CHO remains after 200 hours?

Example 16-8:   Refer to Example 16-7.

(a)   What is the half-life of CH_3CHO if 0.10 mole of it is injected into a 10.0-L vessel at 527°C?

(b)   How many moles of CH_3CHO remain after 200 hours?

(c)   What percent of the CH_3CHO remains after 200 hours?

Let us summarize the results from the last two examples.

	Initial moles CH_3CHO	$[CH_3CHO]_0$	$[CH_3CHO]$ after 200 hrs	Moles of CH_3CHO at 200 hrs	% CH_3CHO Remaining
Ex. 16-7	0.10 mol	0.10 M	0.071 M	0.071 mole	_____%
Ex. 16-8	0.10 mol	0.010 M	0.0096 M	0.096 mole	_____%

Clearly the initial concentration of a reactant influences the rate of reaction *in second order reactions*.

Zero-Order Reaction

For reactions that is *zero order*, the reaction rate is independent of concentrations. The integrated rate equation is

and the half-life is

Study Table 16-2.
ENRICHMENT   Calculus Derivation of Integrated Rate Equations

The deriviation of the integrated rate equations is an example of the use of calculus in chemistry. If you do not know calculus you can still use the results of these derivations. In what follows the integrated rate equations for reactions that are first and second order overall will be derived. For the reaction

$$aA \rightarrow \text{products}$$

the rate is expressed as
$$\text{rate} = - \frac{1}{a}\left(\frac{\Delta[A]}{\Delta t}\right)$$

For a first-order reaction, the rate is proportional to the first power of [A].

$$-\frac{1}{a}\left(\frac{\Delta[A]}{\Delta t}\right) = k[A]$$

In calculus terms, the rate is expressed in terms of an infinitesimal change of concentration $d[A]$ in an infinitesimally short time dt as the derivative of [A] with respect to time.

$$-\frac{1}{a}\left(\frac{d[A]}{dt}\right) = k[A]$$

Separating the variables, (so that we have on one side of the equation only terms in [A] and on the other side of the equation only terms in the other variable t).

$$-\frac{d[A]}{[A]} = (ak)dt$$

We now express this equation in integral form

$$-\int_{[A]_0}^{[A]} \frac{d[A]}{[A]} = (ak)\int_0^t dt$$

250

The equation is to be integrated from time 0 to time t, where the value of [A] at time 0 (the initial value) is $[A]_0$, and the value of [A] at time t is $[A]_t$ or just [A]. This yields

$$-\ln[A] \Big|_0^t = akt \Big|_0^t \quad \text{or}$$

$$-\ln[A]_t + \ln[A]_0 = (ak)t - (ak)o$$

or more simply

$$-\ln[A]_t + \ln[A]_0 = akt$$

which can be rearranged to

$$\ln\frac{[A]_0}{[A]} = akt \quad \text{(first order)}$$

This is the integrated rate equation for a reaction that is first order in reactant A and is first order overall.

For a reaction that is second order in reactant A and second order overall, the rate equation is

$$-\frac{d[A]}{adt} = k[A]^2$$

separating the variables gives

$$\int_{[A]_0}^{[A]} -\frac{d[A]}{[A]^2} = \int_0^t akdt$$

which upon integration gives

$$\frac{1}{[A]} - \frac{1}{[A]_0} = akt$$

The integrated rate equation for a reaction that is second order in reactant A and second order overall.

A zero order reaction has as its rate expression

$$-\frac{d[A]}{adt} = k$$

separating the variables gives

$$\int_{[A]_0}^{[A]} d[A] = \int_0^t -akdt$$

which upon integration gives

$$[A] - [A]_o = -akt$$

that rearranges to the zeroeth order integrated rate law.

$$[A] = [A]_o - akt$$

ENRICHMENT   **Using Integrated Rate Equations To Determine Reaction Order**

We can use the integrated rate equation to analyze concentration-versus-time data. We often plot the data (a graphical approach).

We rearrange the **first-order integrated rate equation**
(recall that $\ln (x/y) = \ln x - \ln y$).

The equation for a straight line is

The first-order integrated rate equation may be interpreted as

$y =$

$m =$

$x =$

$b =$

We see that a plot of $\ln [A]$ versus time should give a straight line with a slope of $-ak$. Study Figure 16-5 carefully.

We show data that can be plotted in three different ways in the next example. Plot these data and determine the order of the reaction. From the linear relationship, calculate the slope of the line and determine the value of the rate constant.

Example 16-9:   Concentration-versus-time data for the thermal decomposition of ethyl bromide are given in the table below. Use the graphs to determine the order of the reaction and the value of the rate constant.

$$C_2H_5Br\,(g) \;\rightarrow\; C_2H_4\,(g) \;+\; HBr\,(g) \qquad\qquad \text{at 700 K}$$

t(min)	0	1	2	3	4	5
$[C_2H_5Br]$	1.00	0.82	0.67	0.55	0.45	0.37
$\ln[C_2H_5Br]$	0.00	-0.20	-0.40	-0.60	-0.80	-0.99
$1/[C_2H_5Br]$	1.00	1.22	1.49	1.82	2.22	2.70

We can determine the value of the rate constant from the slope of the line on the graph of $\ln[C_2H_5Br]$ vs. time.

The integrated rate equation for a reaction that is **second order in reactant A and second order overall**

can be rearranged to

Comparing this equation with the equation for a straight line, we see that

y =

m =

x =

b =

Thus a plot of 1/[A] versus time should give a straight line with slope = _____
and intercept = _____.

Example 16-10:   Concentration-versus-time data for the decomposition of nitrogen dioxide
                   are given in the table below. Use the graphs to determine the order of the
                   reaction and the value of the rate constant.

$$2NO_2 (g) \quad \rightarrow \quad 2NO(g) \quad + \quad O_2(g) \qquad at\ 500\ K$$

t(min)	0	1	2	3	4	5
$[NO_2]$	1.00	0.53	0.36	0.27	0.22	0.18
$\ln[NO_2]$	0.00	-0.63	-1.02	-1.31	-1.51	-1.71
$1/[NO_2]$	1.00	1.89	2.78	3.70	4.55	5.56

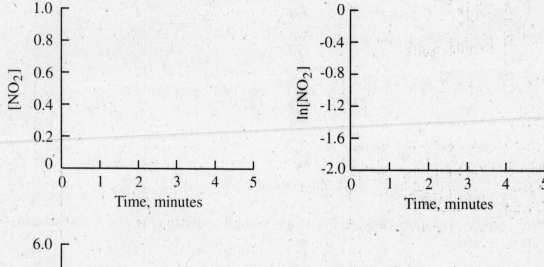

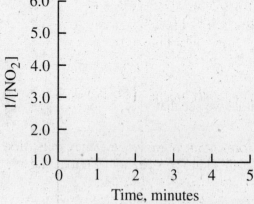

We can determine the value of the rate constant from the slope of the line on the graph of
$1/[NO_2]$ vs. time.

Summary of Concentration Effects

The relationships we discussed in Sections 16-3, 16-4, and 16-5 are summarized in Table 16-2. Study this table carefully.

Summary of Relationships for Various Orders of the Reaction $aA \rightarrow$ Products

	Order		
	Zero	First	Second
Rate-law expression	Rate = k	Rate = k[A]	Rate = k[A]^2
Integrated rate	$[A] = [A]_o - a\mathrm{kt}$	$\ln \dfrac{[A]_o}{[A]} = a\mathrm{kt}$	$\dfrac{1}{[A]} - \dfrac{1}{[A]_o} = a\mathrm{kt}$
Half-life, t$_{1/2}$	$\dfrac{[A]_o}{2ak}$	$\dfrac{\ln 2}{ak} = \dfrac{0.693}{ak}$	$\dfrac{1}{ak[A]_o}$
Plot that gives straight line	[A] vs. t	ln [A] vs. t	$\dfrac{1}{[A]}$ vs. t
Direction of straight-line slope	down with time	down with time	up with time
Interpretation of slope	$-ak$	$-ak$ (for ln plot)	ak
Interpretation of y-intercept	$[A]_0$	$\ln [A]_0$	$\dfrac{1}{[A]_0}$

16-5 Collision Theory of Reaction Rates

The basic idea is: For reactions to occur between atoms, ions, or molecules, they must first collide. Not all collisions are effective collisions. For a collision to be *effective*, the reacting species must

1. possess at least a certain minimum energy necessary to rearrange outer electrons in breaking bonds and forming new ones, and

2. have the proper orientations toward each other at the time of collision.

255

The average kinetic energy of a collection of gaseous molecules is directly proportional to the absolute temperature. See Figure 12-9. At higher temperatures more molecules possess sufficient energy to react.

The rates of reactions that occur by simple *one-step mechanisms* can be explained in terms of "collision theory".

Consider the reaction of methane, the principal component of natural gas, with oxygen. The reaction is exothermic, i.e., it liberates heat.

$$CH_4\,(g) + 2\,O_2\,(g) \rightarrow CO_2\,(g) + 2\,H_2O\,(\ell) + 891\,kJ$$

Opening a gas jet and allowing methane to escape into the air results in no chemical reaction at room temperature. However, striking a match ignites the methane – the reaction is self-sustaining so long as methane and oxygen are mixed in the appropriate proportions *and* the heat produced by the reaction isn't removed too rapidly. What is the function of the lighted match? The heat of the lighted match increases the energy of enough molecules to initiate the reaction of methane with oxygen.

Methane and oxygen do not react at room temperature because the average energies of the CH_4 and O_2 molecules are not high enough to initiate the reaction.

One of the basic ideas of collision theory is that reacting molecules must have the proper orientation with respect to each other when they collide.

$$X_2\,(g) + Y_2\,(g) \rightarrow 2\,XY\,(g)$$

We can visualize collisions between X_2 and Y_2 molecules that are ineffective.

Another kind of collision in which the molecules are properly aligned can be *effective*, i.e., it can result in reaction, *if* the molecules possess sufficient energy (See Figure 16-9).

256

16-6   Transition State Theory

This theory postulates that reactants are converted into a high-energy intermediate state, called a *transition state*, before forming products. See Figure 16-11. In this state the bonds in the reactants are partially broken and those in the products are partially formed. Consider a hypothetical one-step reaction.

$$X_2(g) \quad + \quad Y_2(g) \quad \rightarrow \quad 2XY(g)$$

For X_2 and Y_2 to react, they must possess some minimum energy, called the *activation energy* (Figure 16-10). "Activation energy" may be illustrated by a mechanical analogy.

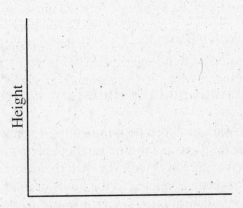

We may indicate the course of the reaction of X_2 and Y_2 graphically.

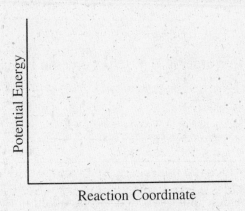

Reaction Coordinate

The relationship between the activation energy for the forward and reverse reactions is

Thus, we see that *collision* of reacting species is a necessary, *but not sufficient,* condition for a chemical reaction to occur. As you observe when a gas jet is opened, the collision of CH_4 and O_2 molecules at room temperature does *not* result in reaction. The distribution of molecules possessing different energies at a given temperature may be represented as a Maxwelliam or Boltzmann distribution.

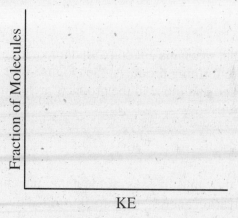

At a higher temperature a larger fraction of molecules possesses sufficient energy to react, the *activation energy,* than at lower temperatures.

16-7   Reaction Mechanisms and the Rate-Law Expression

Once the rate-law expression has been determined for a reaction, it may be used to *postulate a mechanism.* Most reactions occur in a series of elementary or fundametal steps. The step-by step pathway by which a reacion occurs is called the reaction mechanism. The order of a reaction with respect to each reactant is related to the kinds of interactions that occur during, or in some cases preceding, the rate-determining (slowest) step. The *slowest step* in a reaction mechanism is the *rate-determining step.*

For example, consider the iodide ion catalyzed decomposition of hydrogen peroxide to water and oxygen.

The reaction is known to be first order in H_2O_2, first order in I^-, and second order overall. The mechanism is thought to be

Slow step

Fast step _____

Overall reaction.

Note: (1)   one H_2O_2 molecule and *one* I^- ion are involved in the rate-determining step

(2)   the I^- catalyst is consumed in the slow step *and* produced in the fast step in equal amounts

Chemists have identified the hypoiodite ion, IO^-, as a short-lived *reaction intermediate.* This gives us confidence in the proposed mechanism.

In many cases more than one mechanism is consistent with the form of the rate-law expression, and so it is not possible to "prove" that one mechanism is correct. Chemists try to identify *reaction intermediates* to eliminate most possibilities.

Ozone, O_3, reacts very rapidly with nitrogen oxide, NO, in a reaction that is first order in each reactant and second order overall.

$$O_3(g) \; + \; NO(g) \; \rightarrow \; NO_2(g) \; + \; O_2(g) \qquad \text{Rate} = k[O_3][NO]$$

Because the reaction is so rapid it is difficult to identify intermediates. A *possible* mechanism, i.e., a mechanism that is consistent with the rate-law expression, is

Slow step

Fast step _____

Overall reaction

A mechanism that is *inconsistent* with the rate-law expression is

Slow step

Fast step _____

Overall reaction

The experimentally determined reaction orders indicate the number of molecules involved in

1.   the slow step only, *or*

2.   the slow step *and the equilibrium steps* preceding the slow step

16-8   Temperature: The Arrhenius Equation

At a higher temperature, a higher fraction of molecules possesses higher kinetic energies. Increasing the temperature increases reaction rates because more molecules possess energies equal to or greater than the activation energy. Decreasing the temperature has the reverse effect, See Figure 16-13.

From experimental observations. Svante Arrhenius developed the relationship among (1) activation energy, (2) absolute temperature, and (3) the specific rate constant.

or, in logarithmic form,

$e = 2.718$ (the base for natural logarithms)

k is the specific rate constant for the reaction

E_a is the activation energy for the reaction

A is a constant with the same units as the specific rate constant. It is proportional to the frequency of collisions between reacting molecules.

R is the universal gas constant with the same units in its numerator as are used for E_a

T is the absolute temperature

Note that larger values of E_a correspond to smaller values of k, and therefore to slower rates of reaction (other factors being equal).

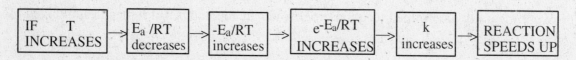

To illustrate *the effect of temperature* on a reaction, let us do the following. For a given reaction, in which the reactant concentrations and E_a are constant, we can develop the following relationship.

1.   Write the Arrhenius equations for two temperatures ($T_2 > T_1$).

2.   Subtract one equation from the other.

3.   Rearrange and solve for $\ln \dfrac{k_2}{k_1}$.

Consider the rate of a reaction, in which $E_a = 50$ kJ/mol, at 20°C and at 30°C.

For a reaction with $E_a \approx 50$ kJ/mol the rate *approximately* doubles for a 10°C rise in temperature, near *room temperature*. Note: two restrictions apply, (1) $E_a \approx 50$ kJ/mol, and (2) temperature is near room temperature.

Consider: $2ICl(g)$ + $H_2(g)$ $\rightarrow$ $I_2(g)$ + $2HCl(g)$

The rate-law expression is known to be $R = k[ICl][H_2]$

At 230°C,

At 240°C,

16-9 Catalysts

Catalysts are substances that speed up reaction rates without being consumed. Most catalysts provide an alternate reaction pathway with a lower activation energy. See Figures 16-15 and 16-16.

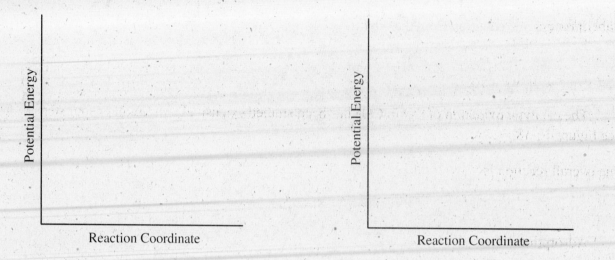

The following shows how the rate of a reaction depends on the value of E_a at *constant temperature* and *the same concentrations* of reactants.

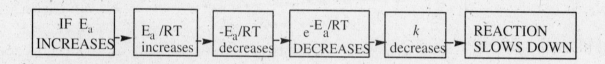

IF E_a INCREASES → E_a/RT increases → $-E_a/RT$ decreases → $e^{-E_a/RT}$ DECREASES → k decreases → REACTION SLOWS DOWN

Homogeneous Catalysis

Homogeneous catalysts exist in the same phase (solid, liquid, or gas) as the reactants.

Heterogeneous Catalysis

Heterogeneous catalysts exist in a different phase than the reactants. Many heterogeneous catalysts are solids. See Figure 16-17. Examples include

Catalytic Convertor Chemistry

Sulfuric acid preparation

Haber process

The catalytic oxidation of CO to CO_2 has been studied extensively. It involves four steps. See Figure 16-18.

The overall reaction is

1.   Adsorption

2.   Activation

3.   Reaction

4.   Desorption

Enzymes as Biological Catalysts

Enzymes are proteins that act as catalysts in living systems. Thousands of enzymes function as catalysts in our bodies.

Synthesis Question

The Chernobyl nuclear reactor accident occurred in 1986. At the time that the reactor exploded some 2.4 MCi of radioactive ^{137}Cs was released into the atmosphere. The half-life of ^{137}Cs is 30.1 years. In what year will the amount of ^{137}Cs released from Chernobyl finally decrease to 100 Ci? A Ci is a unit of radioactivity called the Curie.

Group Question

^{99m}Tc has a half-life of 6.02 hours and is often used in nuclear medical diagnostic tests. Patients are injected with approximately 10 µCi of ^{99m}Tc that is then directed to specific sites in the patient's body to detect gallstones, brain tumors, brain function, and other medical conditions. How long will the patient have a higher than normal radioactivity level after they have been injected with 10 µCi of ^{99m}Tc?

Chapter Seventeen

CHEMICAL EQUILIBRIUM

17-1   Basic Concepts

Most chemical reactions do not go to completion, i.e., all of the reactants are not converted to products. Reversible reactions do not go to completion and can occur in either direction. Reversible reactions may be represented in general terms as

$$a\ A(g)\ +\ b\ B(g)\ \rightleftharpoons\ c\ C(g)\ +\ d\ D(g)$$

where the lower case letters represent stoichiometric coefficients and the capital letters represent formulas. The double arrow indicates that the reaction is reversible, i.e., A and B can react to form C and D, and C and D can react to form A and B. *Chemical equilibrium* exists when the forward and reverse reactions occur *simultaneously at the same rate*. Chemical equilibria are *dynamic equilibria*; that is, individual molecules are continually reacting, even though the overall composition of the reaction mixture does not change.

For the generalized reversible reaction, $A(g) + B(g) \rightleftharpoons C(g) + D(g)$, the rates of the forward and reverse reactions may be represented graphically. Suppose we start with 1.00 mol each of A and B in a 1.00-L container. All reactants and products are understood to be gases unless otherwise indicated. See Figure 17-1.

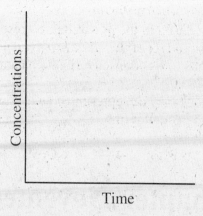

The change in concentration of A and B decreases with time while the change in concentration of C and D increases with time until the rates of the change in concentration of A and B and the rates of change in concentration of C and D are equal. (We often say that the rate of the forward (left-to-right) reaction equals the rate of the reverse (right-to-left) reaction. At that point *equilibrium is established.*

Many experiments on reversible reactions have demonstrated the validity of a very fundamental idea: It makes no difference whether we start with the "reactants" or the "products" of a reversible reaction, because equilibrium can be established from either direction.

17-2 The Equilibrium Constant

Consider a reversible reaction that occurs by a *simple one-step mechanism* in both directions.

$$aA(g) \; + \; bB(g) \; \rightleftharpoons \; cC(g) \; + \; dD(g)$$

The rate of the forward reaction can be represented as

> The square brackets indicate *equilibrium* concentrations in moles/liter.

The rate of the reverse reaction can be represented as

When the rates of the forward and reverse reactions are equal, the system is at equilibrium. From the definition of equilibrium we can write

The **equilibrium constant, K$_c$**, is defined as the product of the *equilibrium concentrations* (moles/liter) of the products, each raised to the power that corresponds to its coefficient in the balanced equation, divided by the product of the *equilibrium concentrations* of reactants, each raised to the power that corresponds to its coefficient in the balanced equation.

In *general* terms reversible reactions may be represented as follows. Equilibrium constant expressions may also be written in general terms.

$$a A(g) \quad + \quad b B(g) \quad \rightleftharpoons \quad c C(g) \quad + \quad d D(g)$$

We have described a *one-step* reaction. Similar reasoning leads to the same result for a *two-step* reaction.

Equilibrium constants are thermodynamic quantities. They vary with temperature only. Equilibrium constants must be calculated from experimental data. They can be calculated if equilibrium concentrations of all reactants and products are known, or they can be calculated from thermodynamic data, as we shall see shortly.

Example 17-1:   Write equilibrium constant expressions for the following reactions at 500°C. All reactants and products are *gases* at 500°C.

$$PCl_5(g) \quad \rightleftharpoons \quad PCl_3(g) \quad + \quad Cl_2(g)$$

$$H_2(g) \quad + \quad I_2(g) \quad \rightleftharpoons \quad 2 HI(g)$$

$$4 NH_3(g) \quad + \quad 5 O_2(g) \quad \rightleftharpoons \quad 4 NO(g) \quad + \quad 6 H_2O(g)$$

The actual thermodynamic definition of the equilibrium constant involves *activities* rather than concentrations. The *activity* of a component of an *ideal* mixture is the ratio of its concentration or partial pressure to a standard concentration (1.00 *M*) or partial pressure (1.00 atm). For now we can consider the activity of each species to be a *dimensionless* quantity whose numerical value may be determined as follows:

1. For *pure* solids and liquids, the activity is taken as 1.

2. For components of ideal solutions, the activity of each component is taken to be equal to its molar concentration.

3. For mixtures of ideal gases, the activity of each component is taken to be equal to its partial pressure in atmospheres.

Because we use activities, *equilibrium constants have no units*. Let us illustrate how equilibrium constants are determined experimentally.

The magnitude of K_c, is a measure of the extent to which reaction occurs. For any balanced chemical equation, the value of K_c

   1. is constant at a given temperature

   2. changes if the temeperature changes

   3. does not depend on the initial concentrations

Example 17-2:   One liter of equilibrium mixture from the following reaction system at a high temperature was found to contain 0.172 mole of phosphorus trichloride, 0.086 mole of chlorine, and 0.028 mole of phosphorus pentachloride. Calculate K_c for the reaction.

$$PCl_5(g) \rightleftharpoons PCl_3(g) + Cl_2(g)$$

Equil. []s

Example 17-3:   The decomposition of PCl_5 was studied at another temperature. One mole of PCl_5 was introduced into an evacuated 1.00-liter container. The system was allowed to reach equilibrium at the new temperature. At equilibrium 0.60 mole of PCl_3 was present in the container. Calculate the equilibrium constant at this temperature.

$$PCl_5(g) \rightleftharpoons PCl_3(g) + Cl_2(g)$$

Initial

Change
Equilibrium
_____

Example 17-4:   At a given temperature 0.80 mole of N_2 and 0.90 mole of H_2 were placed in an evacuated 1.00-liter container. At equilibrium 0.20 mole of NH_3 was present. Calculate K_C for the reaction.

$$N_2(g) \quad + \quad 3\,H_2(g) \quad \rightleftharpoons \quad 2\,NH_3(g)$$

Initial

Change
_____

Equilibrium

17-3 Variation of K_C with the Form of the Balanced Equation

The value of K_C depends upon the form of the balanced equation. In Example 17-2 we wrote the equation for the decomposition of PCl_5 and its equilibrium constant as

$$PCl_5(g) \quad \rightleftharpoons \quad PCl_3(g) \quad + \quad Cl_2(g) \qquad K_C = \frac{[PCl_3][Cl_2]}{[PCl_5]} = 0.53$$

Example 17-5:   Calculate the equilibrium constant for the reverse reaction by two methods, i.e., the equilibrium constant for the reaction,

$$PCl_3(g) \quad + \quad Cl_2(g) \quad \rightleftharpoons \quad PCl_5(g)$$

Use the
Equilibrium
From Ex. 17-2

271

We see that: $K_c = 1/K_c'$ or $K_c' = 1/K_c$

This relationship between K_c and K_c' is valid for all equilibrium constant expressions.

An equilibrium constant indicates the extent to which a reaction can occur. A large equilibrium constant tells us that most of the reactants are converted into products before equilibrium is established. Conversely, a small equilibrium constant indicates that most of the reactants remain, and that only small amounts of products are formed at equilibrium.

17-4 The Reaction Quotient

The *reaction quotient*, Q, has the same form as the equilibrium constant, K_c, but the concentrations are ***not necessarily equilibrium concentrations***.

Comparison of Q with K_c enables us to predict the direction in which a reaction will occur to a greater extent when a system is *not* at equilibrium. The relationship between Q and K_c as well as the direction in which a reaction occurs to the greater extent in a system.*not* at equilibrium, may be summarized.

When   Q = K   the system is at equilibrium

When   Q > K   the reaction occurs to the _____ to a greater extent

When   Q < K   the reaction occurs to the _____ to a greater extent

Example 17-6:   The equilibrium constant for the following reaction is 49 at 450°C. If 0.22 mole of I_2, 0.22 mole of H_2 and 0.66 mole of HI were put into an evacuated 1.00-liter container at 450°C, would the system be at equilibrium? If not, what must occur to establish equilibrium?

$$H_2(g) \quad + \quad I_2 \quad \rightleftharpoons \quad 2\,HI(g)$$

We calculate Q

272

We see that Q < K, which tells us that the system is not at equilibrium and that the forward reaction occurs to a greater extent until equilibrium is established. We introduce Q here. This quantity will be used in Chapters 17, 28, 19, 20, and 21.

17-5 Uses of the Equilibrium Constant, K_C

Example 17-7:   The equilibrium constant, K_C, is 3.00 for the following reaction at a given temperature. If 1.00 mole of SO_2 and 1.00 mole of NO_2 are put into an evacuated 2.00-liter container and allowed to reach equilibrium, what will be the concentration of each compound at equilibrium?

$$SO_2(g) \quad + \quad NO_2(g) \quad \rightleftharpoons \quad SO_3(g) \quad + \quad NO(g)$$

Initial

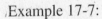Change _____

Equilibrium

Example 17-8:   The equilibrium constant is 49 for the following reaction at 450°C. If 1.00 mole of HI is put into an evacuated 1.00-liter container and allowed to reach equilibrium, what will be the equilibrium concentration of each substance?

$$H_2(g) \quad + \quad I_2(g) \quad \rightleftharpoons \quad 2\,HI(g)$$

Initial

Change _____

Equilibrium

17-6   Disturbing a System at Equilibrium: Predictions

An important scientific principle that affects chemical equilibria is known as *LeChatelier's Principle*. **If a change of conditions (stress) is applied to a system in equilibrium, the system responds in the way that best tends to reduce the stress in reaching a new state of equilibrium.** Changes in concentrations, pressure, and temperature are considered stresses. The effect of each will be illustrated in the next few examples.

It may be convenient to express the amount of a gas present in terms of its partial pressure rather than its concentration. Solving the ideal gas equation for pressure gives

because
n/V has the
units, mol/L

At *constant temperature*, the partial pressure of a gas is directly proportional to its concentration.

Let us examine the factors that affect systems in equilibrium qualitatively and then quantitatively. The four factors are:

1.   **Changes in Concentration** (Pressure for reactions involving gases)

Suppose we have the following system at equilibrium at 450°C ($K_c = 49$), and we

$$H_2(g) + I_2(g) \rightleftharpoons 2\,HI(g)$$

Add some H_2

Remove some H_2

2.   **Changes in Volume and Pressure**

At constant temperature, the pressure and volume of a sample of gas are inversely proportional (Boyle's Law). If we increase the pressure of a system at equilibrium at constant temperature by decreasing the volume, the concentrations of all gases (but not liquids or solids) increase. If a balanced equation has more moles of gaseous reactants than gaseous products, the forward reaction is favored because this tends to relieve the stress. A decrease in pressure caused by an increase in volume has the opposite effect. Increasing the total pressure by adding an inert gas has no effect.

Consider the effect of changing the pressure by changing the volume *at constant temperature* on the following system at equilibrium:

274

$$2 \, NO_2(g) \quad \rightleftharpoons \quad N_2O_4(g)$$

If the volume is decreased
(pressure increased)

If the volume is increased
(pressure decreased)

Increasing the pressure (decreasing the volume) favors the reaction that produces fewer moles of gases, and vice-versa.

3.   **Changes in Temperature**

Consider the following exothermic reaction at equilibrium.

$$2 \, SO_2(g) \quad + \quad O_2(g) \quad \rightleftharpoons \quad 2 \, SO_3(g) \quad + \quad 198 \, kJ$$

Increasing the
temperature

Decreasing the
temperature

Increasing the temperature always favors the reaction that consumes heat, and vice-versa.

4.   **Addition of a Catalyst**

A catalyst decreases the activation energy of both the forward and reverse reactions equally. It does not affect the position of equilibrium, but it decreases the time required to reach equilibrium.

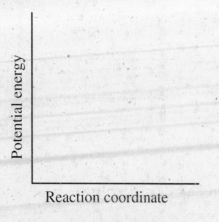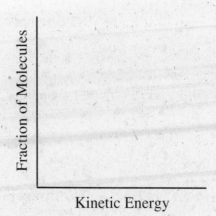

Example 17-9: Given the reaction below at equilibrium in a closed container at 500°C. How would the equilibrium be influenced by each of the following?

$$N_2(g) \quad + \quad 3\,H_2(g) \quad \rightleftharpoons \quad 2\,NH_3(g) \quad + \quad 92\,kJ$$

a. Increasing the temperature _____

b. Lowering the temperature _____

c. Increasing the pressure by decreasing the volume _____

d. Introducing some platinum catalyst _____

e. Forcing more H_2 into the system _____

f. Removing some NH_3 from the system _____

Example 17-10: How will an increase in pressure (caused by decreasing the volume) affect the equilibrium in each of the following reactions?

a. $H_2(g) + I_2(g) \rightleftharpoons 2\,HI(g)$ _____

b. $4\,NH_3(g) + 5\,O_2(g) \rightleftharpoons 4\,NO(g) + 6\,H_2O(g)$ _____

c. $PCl_3(g) + Cl_2(g) \rightleftharpoons PCl_5(g)$ _____

d. $2\,H_2(g) + O_2(g) \rightleftharpoons 2\,H_2O(g)$ _____

Example 17-11:   How will an increase in temperature affect each of the following reactions?

a. $2 NO_2(g) \rightleftharpoons N_2O_4(g) + heat$ _____

b. $H_2(g) + Cl_2(g) \rightleftharpoons 2 HCl(g) + 92 kJ$ _____

c. $H_2(g) + I_2(g) \rightleftharpoons 2HI(g)$ $\Delta H = +25 kJ$ _____

17-7   The Haber Process: A Commercial Application of Equilibrium

The Haber Process for the commercial production of ammonia is a classic example of the application of the concepts we have evolved to an extremely important reaction.

Table 17-1
Effect of T and P on Yield of Ammonia

°C	K_c	MOLE % NH_3 IN EQUILIBRIUM MIXTURE		
		10 atm	100 atm	1000 atm
209	650	51	82	98
467	0.5	4	25	80
758	0.014	0.5	5	13

17-8   Disturbing a System at Equilibrium: Calculations

We use equilibrium constants to determine the *quantitative* changes that occur when a stress is applied to a system in equilibrium.

Concentration changes (pressure changes in gas phase reactions) shift equilibria. To determine the direction in which concentration changes (pressure changes for gas phase reactions) shift an equilibrium, we calculate the reaction quotient, Q, and compare it to K_c.

Example 17-12:   An equilibrium mixture from the following reaction was found to contain 0.20 mol/L of A, 0.30 mol/L of B, and 0.30 mol/L of C. What is the value of K_c?

$$A(g) \rightleftharpoons B(g) + C(g)$$

Equil. []'s

If the volume of the reaction vessel were suddenly <u>doubled</u> while the temperature remained constant, what would be the new equilibrium concentrations? First, let us calculate Q, the reaction quotient, *after the volume has been doubled*.

$$A(g) \rightleftharpoons B(g) + C(g)$$

Instantaneous []'s

We see that $Q < K_C$. This tells us that the reaction to the right occurs to a greater extent to re-establish equilibrium. We represent the new equilibrium concentrations algebraically and then calculate them.

$$A(g) \rightleftharpoons B(g) + C(g)$$

New Initial []

<u>Change</u>

New Equilibrium []

Example 17-13: Refer to Example 17-12. If the initial volume of the reaction vessel were <u>halved</u>, while the temperature remains constant, what will the new equilibrium concentrations be? Recall: [A] = 0.20 *M*, [B] = 0.30 *M*, and [C] = 0.30 *M originally*.

$$A(g) \rightleftharpoons B(g) + C(g)$$

Instantaneous []'s

First, we calculate Q *after the volume has been halved*.

$Q > K$, so the reaction to the left occurs to the greater extent. The algebraic representations of the new equilibrium concentrations are

$$A(g) \rightleftharpoons B(g) + C(g)$$

New Initial []

Change _____

New Equilibrium []

Example 17-14:   A 2.00-liter vessel in which the following system is at equilibrium contains 1.20 moles of $COCl_2$, 0.60 mole of CO and 0.20 mole of Cl_2. Calculate the equilibrium constant.

$$CO(g) + Cl_2(g) \rightleftharpoons COCl_2(g)$$

Equil. []'s

An additional 0.80 mole of Cl_2 is added to the vessel at the same temperature. Calculate the molar concentrations of CO, Cl_2, and $COCl_2$ when the new equilibrium is established.

$$CO(g) + Cl_2(g) \rightleftharpoons COCl_2(g)$$

Original Equilibrium []

(Stress Add) _____
New Initial []

Change _____
New Equilibrium []

17-9 Partial Pressures and the Equilibrium Constant

For gas phase equilibria, equilibrium constants may be expressed in terms of partial pressures as well as in terms of concentrations. Recall that the volume of a gas varies inversely with pressure, and therefore the concentration of a gas is directly proportional to the applied pressure. Consider the following system in equilibrium at 500°C.

$$2\ Cl_2(g)\ +\ 2\ H_2O(g)\ \rightleftharpoons\ 4HCl(g)\ +\ O_2(g)$$

17-10 Relationship Between K_p and K_c

The relationship between an equilibrium constant expressed in terms of concentrations, K_c, and the equilibrium constant expressed in terms of partial pressures, K_p, is

Δn = (# of moles of gaseous products) - (# of moles of gaseous reactants)

Example 17-15:   Nitrosyl bromide, NOBr, is 34% dissociated by the following reaction at 25°C, in a vessel in which the total pressure is 0.25 atmosphere. What is the value of K_p?

$$2\ NOBr(g)\ \rightleftharpoons\ 2\ NO(g)\ +\ Br_2(g)$$

Initial []'s

Change

Equilibrium []'s

280

The numerical value of K_C for this reaction is

Example 17-16: K_C is 49 for the following reaction at 450°C. If 1.0 mole of H_2 and 1.0 mol of I_2 are allowed to reach equilibrium in a 3.0-liter vessel.

(a)   How many moles of I_2 remain unreacted at equilibrium?

$$H_2(g) \quad + \quad I_2(g) \quad \rightleftharpoons \quad 2\,HI(g)$$

Initial []'s

Change

Equilibrium []'s

(b)   What are the equilibrium partial pressures of H_2, I_2 and HI?

(c)   What is the total pressure in the reaction vessel?

281

17-11   Heterogeneous Equilibria

Heterogeneous equilibria involve two or more phases. The *activities* of pure solids and pure liquids are unity. The following examples illustrate how equilibrium constant expressions are written for heterogeneous equilibria. Solvents in dilute solutions are treated as "nearly pure liquids", i.e., their activities are taken as 1.

$$CaCO_3(s) \rightleftharpoons CaO(s) + CO_2(g) \qquad \qquad (\text{at } 500°C)$$

$$SO_2(g) + H_2O(l) \rightleftharpoons H_2SO_3(aq) \qquad \qquad (\text{at } 25°C)$$

$$CaF_2(s) \rightleftharpoons Ca^{2+}(aq) + 2\,F^-(aq) \qquad \qquad (\text{at } 25°C)$$

$$3\,Fe(s) + 4\,H_2O(g) \rightleftharpoons Fe_3O_4(s) + 4\,H_2(g) \qquad (\text{at } 500°C)$$

17-12   Relationship Between ΔG^0_{rxn} and the Equilibrium Constant

The *standard* free energy change for a reaction is ΔG^0_{rxn}. This is the free energy change that would accompany the *complete* conversion of all reactants, initially present in their *standard states*, to *all* products in their *standard* states. ΔG is the free energy change for other (*nonstandard*) and pressures. The relationship between ΔG and $\Delta G°$ is

R is the universal gas constant
T is the absolute temperature
Q is the reaction quotient

When a system is at equilibrium, $\Delta G = 0$ and $Q = K$. Then

Rearranging gives

For the following generalized reaction, the *thermodynamic equilibrium* constant is defined as,

$$a\,A \;+\; b\,B \;\rightleftharpoons\; c\,C \;+\; d\,D$$

a_A is the activity of A
a_B is the activity of B
a_C is the activity of C
a_D is the activity of D

The mass action expression to which the thermodynamic equilibrium constant is related involves concentrations for species in solution and partial pressures for gases.

For reactions that involve gases, the *thermodynamic equilibrium constant* is K_P. For reactions that involve species in solution, it is K_C.

The relationships among ΔG^0_{rxn}, K, and the spontaneity of a reaction (initial concentrations of 1.00 M, or partial pressures of 1.00 atmosphere for gases) follow.

ΔG^0_{rxn} is   , K > 1   Forward reaction would be spontaneous at unit concentrations or partial pressures.

ΔG^0_{rxn} is   , K = 1   System would be at equilibrium at unit concentrations or partial pressure (very rare case)

ΔG^0_{rxn} is   , K < 1   Reverse reaction would be spontaneous at unit concentrations or partial pressures

Example 17-17:   Calculate the equilibrium constant, K_P, for the following reaction at 25°C from the thermodynamic data in Appendix K.

$$N_2O_4(g) \;\rightleftharpoons\; 2\,NO_2(g)$$

K_p for the reverse reaction at 25°C can be calculated easily, because it is the reciprocal of that for the above reaction.

$$2\ NO_2(g) \rightleftharpoons N_2O_4(g) \quad \Delta G^0_{rxn} = -4.78\ kJ/mol\ (spontaneous\ rxn)$$

Example 17-18:   At 25°C and 1.00 atmosphere, $K_p = 4.3 \times 10^{-13}$ for the decomposition of NO_2. Calculate ΔG^0_{rxn} at 25°C.

$$2\ NO_2(g) \rightleftharpoons 2\ NO(g) + O_2(g)$$

At conditions *other than thermodynamic standard state conditions*, we must use the following relationship to calculate equilibrium constants.

This means that we must first know G or ΔG^0_{rxn} (or calculate them from tabulated data). The Q term includes concentrations and partial pressures.

17-13   Evaluation of Equilibrium Constants at Different Temperatures

If we know $\Delta H°$ and the equilibrium constant for a reaction at one temperature (T_1), the *van't Hoff equation* allows us to estimate the equilibrium constant for the reaction at another temperature (T_2). Temperatures must be expressed in K.

$$\ln\left(\frac{K_{T_2}}{K_{T_1}}\right) = \frac{\Delta H°(T_2 - T_1)}{RT_2T_1} \quad \text{or} \quad \log\left(\frac{K_{T_2}}{K_{T_1}}\right) = \frac{\Delta H°\ (T_2 - T_1)}{2.303\ RT_2T_1}$$

Example 17-19:   For the reaction in Example 17-18, DH° = 114 kJ/mol and $K_p = 4.3 \times 10^{-13}$ at 25°C. Estimate K_p at 250°C

$$2\ NO_2(g) \rightleftharpoons 2\ NO(g) + O_2(g)$$

In this example we make $T_1 = 298$ K, and $T_2 = 523$ K. We then evaluate the right side of the van't Hoff equation

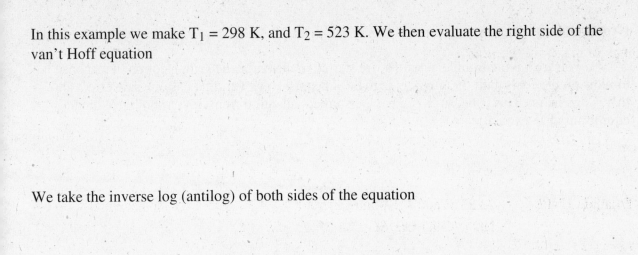

We take the inverse log (antilog) of both sides of the equation

Now we solve for K_{T_2} and substitute the known value of K_{T_1}

Synthesis Question

Mars is a reddish colored planet because it has numerous iron oxides in its soil. Mars also has a very thin atmosphere, although it is believed that quite some time ago its atmosphere was considerably thicker. The thin atmosphere does not retain heat well, thus at night on Mars the surface temperatures are 145 K and in the daytime the temperature rises to 300 K. Does Mars get redder in the daytime or at night?

Group Question

If you are having trouble getting a fire started in the barbecue grill, a common response is to blow on the coals until the fire begins to burn better. However, this has one side effect and that is dizziness. This is because you have disturbed an equilibrium in your body. Which equilibrium have you affected?

Chapter Eighteen

IONIC EQUILIBRIA I: ACIDS AND BASES

Earlier we classified compounds as electrolytes or as nonelectrolytes. *Electrolytes* are compounds whose aqueous solutions conduct electric current because they ionize or dissociate in aqueous solution. *Nonelectrolytes* are compounds whose aqueous solutions do not conduct electric current because they do not ionize or dissociate in aqueous solutions.

18-1   A Review of Strong Electrolytes

Strong electrolytes ionize or dissociate completely, or very nearly completely, in dilute aqueous solutions. Three common classes of compounds are strong electrolytes: (1) strong acids, (2) strong soluble bases, and (3) most soluble salts. Refer to Section 4-2, part 5 for the solubility rules.

1.   strong acids

2.   strong bases

3.   soluble salts

We can calculate concentrations of ions in solutions of strong electrolytes easily.

Example 18-1:   Calculate the concentrations of ions in 0.050 M nitric acid, HNO_3, solution.

Thus, 0.050 M HNO_3 solution contains H_3O^+ and NO_3^- ions in equal concentrations, i.e., each is 0.050 M, and essentially *no molecules* of nonionized nitric acid are present.

Example 18-2:   Calculate the concentrations of ions in 0.020 M calcium hydroxide, $Ca(OH)_2$, solution.

Thus, 0.020 M calcium hydroxide soltuion contains 0.020 M Ca^{2+} ions and 0.040 M OH^- ions, and essentially *no molecules* of unionized calcium hydroxide.

18-2   The Autoionization of Water

Careful experiments have shown that pure water ionizes very slightly. The activity of pure H_2O is 1, so the equilibrium constant for this reaction is

Careful measurements show that the concentration of each ion is 1.0×10^{-7} mol/L at 25°C.

This equilibrium constant is called the **ion product for water, K_w**.

The K_w expression was obtained for pure water. This relationship is valid for all dilute aqueous solutions at 25°C. You should remember it. Like other equilibrium constants, K_w varies only with temperature. See Table 18-2.

Example 18-3:   Calculate the concentrations of H_3O^+ and OH^- ions in 0.050 M HCl.

In solving this problem, we have assumed that *all* the H_3O^+ ions come from the ionization of HCl, i.e., we neglected the H_3O^+ ions formed by the ionization of water. This is a valid assumption because the ionization of *pure* water produces only 1.0×10^{-7} M H_3O^+. When HCl is dissolved in water, large numbers of H_3O^+ are produced. The large increase in H_3O^+ concentration from HCl shifts the water equilibrium to the left sharply (LeChatelier's Principle), and so the concentration of OH^- ions is decreased dramatically.

In acidic solutions the H_3O^+ concentration is always greater than the OH^- concentration. We must not conclude that acidic solutions contain no OH^-, but rather that the concentration of OH^- is always less than 1.0×10^{-7} mol/L in acidic solutions. The converse is true for basic solutions.

18-3   The pH and pOH Scales

The pH scale provides a convenient way to express the acidity and basicity of dilute aqueous solutions. The pH of a solution is defined as

A number of "p" terms are used. In general, a lower case "p" before a symbol is read "negative logarithm of" the symbol.

If either H_3O^+ or OH^- concentrations are known, pH and pOH can be calculated.

Example 18-4:   Calculate the pH of a solution in which $[H_3O^+] = 0.030\ M$.

Example 18-5:   The pH of a solution is 4.597. What is the concentration of H_3O^+?

A convenient relationship between pH and pOH may be derived for all dilute aqueous solutions at 25°C.

Taking the logarithm of both sides of this equation gives

Multiplying both sides of this equation by -1 gives

or, in slightly different form, we have

292

We now have expressions relating [H$_3$O$^+$] and [OH$^-$] as well as pH and pOH at 25°C. Both of these expressions should be remembered.

$$[H_3O^+][OH^-] = 1.0 \times 10^{-14} \quad \text{and} \quad pH + pOH = 14.00$$

The usual range for
the pH scale is:

Example 18-6:   Calculate [H$_3$O$^+$], pH, [OH$^-$], and pOH for a 0.020 M HNO$_3$ solution.

To develop familiarity with the pH scale, consider a series of solutions in which the H$_3$O$^+$ concentration varies between 1.0 M and 1.0 $\times$ 10^{-14} M. Study Table 18-3 carefully.

[H$_3$O$^+$] (M)	[OH$^-$] (M)	pH	pOH
1.0			
1.0 $\times$ 10^{-3}			
1.0 $\times$ 10^{-7}			
2.0 $\times$ 10^{-12}			
1.0 $\times$ 10^{-14}			

Example 18-7:   Calculate the number of H$_3$O$^+$ and OH$^-$ ions in one liter of pure water at 25°C.

The pH of a solution may be determined by using indicators (Figure 18-2), or by using a pH meter (Figure 18-1). The pH meter utilizes a glass electrode, a sensing device that generates a voltage that is proportional to the pH of the solution in which it is immersed. An electronic circuit amplifies the voltage generated by the glass electrode. The pH of the solution is read directly from a scale.

18-4 Ionization Constants for Weak Monoprotic Acids and Bases

Thus far we have discussed strong acids and strong soluble bases. Weak acids and weak bases, *by definition*, ionize only slightly in dilute aqueous solution. Let us consider the reaction that occurs when a weak acid, such as acetic acid, is dissolved in water. The equation for the ionization of acetic acid, CH_3COOH, is

We write the equilibrium constant expression for this reaction

This expression contains the concentration of water, which is very high in dilute aqueous solutions. One liter of pure water contains 55.5 moles of water. Thus, in dilute aqueous solutions, for all practical purposes,

Because the concentration of water is essentially constant in dilute aqueous solutions (its *activity* is taken as one), we can rearrange the above equilibrium constant expression to

This is the expression for the *ionization constant*, K_a, for acetic acid.

In *simplified* form this equation and ionization constant expression are written as

Like other equilibrium constants, ionization constants must be determined experimentally. Values for several ionization constants are listed below. See Appendix F for others.

Ionization Constants for Some Weak Acids (25°C)

Acid	Formula	Ionization Constant
acetic acid	CH_3COOH	1.8×10^{-5}
nitrous acid	HNO_2	4.5×10^{-4}
hydrofluoric acid	HF	7.2×10^{-4}
hypochlorous acid	HOCl	3.5×10^{-8}
hydrocyanic acid	HCN	4.0×10^{-10}

Ionization constants are equilibrium constants for ionization reactions and so they indicate the extent to which weak acids ionize. Acids with large ionization constants ionize to a greater extent than acids with smaller ionization constants. Reference to the above table shows that the order of increasing acid strengths for these five weak acids is

The order of increasing base strength of the anions (conjugate bases) of these acids is

Example 18-8: Write the equation for the ionization of the weak acid HCN and the expression for its ionization constant.

Calculation of Ionization Constants

    Ionization constants may be determined by a number of methods. For example, freezing point depression measurements, electrical conductivity measurements, and pH measurements enable us to calculate ionization constants for weak acids and weak bases.

Example 18-9:   In 0.12 M solution, a weak monoprotic acid, HY, is 5.0% ionized. Calculate the ionization constant for the weak acid.

The ionization equation and the ionization constant expression for HY are

The weak acid is 5.0 percent ionized, so it 95.0% nonionized. Equilibrium concentrations of the various species are

Substitution into the ionization constant expression gives the value for K_a

Example 18-10:   The pH of a 0.10 M solution of a weak monoprotic acid, HA, is found to be 2.97. What is the value for its ionization constant?

The pH of the solution is 2.97, so we can calculate the hydrogen ion concentration.

Recall, pH = -log [H$_3$O$^+$]   or   [H$_3$O$^+$] = 10^{-pH}.

Now that we know the $[H_3O^+]$ in the solution, we can represent the concentrations of the other species and substitute them into the ionization constant expression.

$$HA \rightleftharpoons H^+ + A^-$$

Equilibrium []'s

If the ionization constant and concentration of a weak acid solution are known, the concentrations of the ions in the solution can be calculated.

Calculation of Concentrations from K_a

Example 18-11:   Calculate the concentrations of the various species in 0.15 M
acetic acid, CH_3COOH, solution.

The equation for the ionization of acetic acid and its ionization constant expression are

We represent the equilibrium concentrations of the species in 0.15 M acetic acid solution, and substitute them into the equilibrium constant expression. We let $x\ M$ = mol/L of CH_3COOH that ionize.

$$CH_3COOH \rightleftharpoons H^+ + CH_3COO^-$$

Initial []'s

Change

Equilibrium []'s

Solving the quadratic equation by the quadratic formula gives the same results.

Rule of Thumb: When the linear variable, x, is added to or subtracted from a large number compared to K_a, *and* when $K_a = 10^{-4}$ or less, x may be disregarded. (The "x" represents the "amount of reaction" in the forward direction.)

Let us now calculate the percent ionization for the 0.15 M acetic acid solution. From Example 18-11, we know the concentration of CH_3COOH that ionizes in this solution. The percent ionization of acetic acid in 0.15 M acetic acid is.

Example 18-12: Calculate the concentrations of the species in 0.15 M hydrocyanic acid, HCN, solution.

We represent the concentrations of various species algebraically. We let x M = mol/L of HCN that ionize. Substitution into the K_a expression and making the simplifying assumption we made earlier gives

$$HCN \rightleftharpoons H^+ + CN^-$$

Initial []'s

Change
_____

Equilibrium []'s

The percent ionization of 0.15 M HCN solution is calculated as in the previous example.

To gain some appreciation for the extent of ionization of weak monoprotic acids by looking at their ionization constants, let us compare 0.15 M CH$_3$COOH and 0.15 M HCN solution, i.e., Examples 18-11 and 18-12.

Solution	K_a	[H$^+$]	pH	% ionization
0.15 M CH$_3$COOH	1.8×10^{-5}	_____ M	2.80	_____
0.15 M HCN	4.0×10^{-10}	_____ M	5.11	_____

We see that [H$^+$] in 0.15 M CH$_3$COOH solution is approximately _____ times greater than in 0.15 M HCN solution.

To this point we have discussed weak acids. Aqueous ammonia and its organic derivatives, the amines, are the common soluble weak bases. They are treated in the same way as solutions of weak acids. See Table 18-6 and Appendix G.

Example 18-13:   Calculate the concentrations of the various species in 0.15 M aqueous ammonia solution.

The equation for the ionization of ammonia in water, its ionization constant expression, and the algebraic representations of concentrations of the various species follow. We let x M = [NH$_3$] in mol/L that ionize.

$$NH_3 \ + \ H_2O \ \rightleftharpoons \ NH_4^+ \ + \ OH^-$$

Initial []'s

Change _____

Equilibrium []'s

We calculate the percent ionization of a weak base like we did for a weak acid.

The ionization constant for aqueous ammonia is equal to the ionization constant for acetic acid. This tells us that in aqueous solutions of the *same* concentration NH_3 and CH_3COOH ionize to the same extent. This is to be expected because the K_a for acetic acid and the K_b for ammonia are both 1.8×10^{-5}.

Example 18-14:   The pH of an aqueous ammonia solution is 11.37. Calculate the molarity (original concentration) of the aqueous ammonia solution.

The equation for the ionization of aqueous NH_3 gives us the representation of the equilibrium concentrations.

$$NH_3 \;+\; H_2O \;\rightleftharpoons\; NH_4^+ \;+\; OH^-$$

Initial []'s

Change
_____

Equilibrium []'s

Substitution into the ionization constant expression gives

Examination of the equation suggests that $x - 2.3 \times 10^{-3} \approx x$. Making this assumption simplifies the calculation and gives

18-5   Polyprotic Acids

Thus far, we have discussed weak monoprotic acids, i.e., acids that contain only one acidic hydrogen per formula unit. Many weak acids contain two or more acidic hydrogens. These are called polyprotic acids. Polyprotic acids ionize stepwise. An ionization constant can be written for each step in the ionization.

Consider arsenic acid, H_3AsO_4, as a typical polyprotic acid. The ionization constants are $K_{a1} = 2.5 \times 10^{-4}$, $K_{a2} = 5.6 \times 10^{-8}$, $K_{a3} = 3.0 \times 10^{-13}$. The first ionization step is

The second ionization step is

The third ionization step is

Examination of the three ionization constants for arsenic acid shows that

This is generally true for inorganic polyprotic acids. Refer to Appendix F. Successive ionization constants usually decrease by a factor of approximately 10^4 to 10^6. Large decreases in the values of successive ionization constants tell us that each step in the ionization of a polyprotic acid occurs to a much lesser extent than the previous step.

Example 18-15:   Calculate the concentrations of all species in 0.100 M arsenic acid, H_3AsO_4, solution.

We write the equation for the first step ionization and represent the concentrations

We substitute into the expression for K_{a1}.

$$K_{a1} = \frac{[H_3O^+][H_2AsO_4^-]}{[H_3AsO_4]} = 2.5 \times 10^{-4}$$

We solve this equation by the quadratic formula, and obtain two values for x

Now we write the equation for the second step ionization and represent the concentrations.

Substitution into the expression for the second step ionization constant gives

$$K_{a2} = \frac{[H_3O^+][HAsO_4^{2-}]}{[H_2AsO_4^-]} = 5.6 \times 10^{-8}$$

For solutions of weak polyprotic acids in *reasonable* concentrations for which $K_{a1} \gg K_{a2} \gg K_{a3}$, we always find that

$$[Anion^{2-}] = K_{a2}$$

We write the equation for the third step ionization and represent the concentrations

Substitution into the third step ionization constant expression gives

$$K_{a3} = \frac{[H_3O^+][AsO_4^{3-}]}{[HAsO_4^{2-}]} = 3.0 \times 10^{-13}$$

We use K_w to calculate $[OH^-]$ in the 0.100 M H_3AsO_4 solution.

A comparison of concentrations of the various species in 0.100 M H_3AsO_4 solution follows.

Species	Concentration
H_3AsO_4	_____ M
H^+	_____ M
$H_2AsO_4^-$	_____ M
$HAsO_4^{2-}$	_____ M
AsO_4^{3-}	_____ M
OH^-	_____ M

Note that nonionized H_3AsO_4 is present in greater concentration than any other species in 0.100 M H_3AsO_4 solution. The only other species present in significant concentrations are H^+ and $H_2AsO_4^-$, the species formed in the first step. Similar statements can be made for other weak polyprotic acids for which, $K_{a1} \gg K_{a2} \gg K_{a3}$. See Tables 18-7 and 18-8.

18-6   Solvolysis

Solvolysis describes the reaction of a dissolved substance with the solvent. When solvolysis reactions occur in aqueous solution, they are called *hydrolysis reactions*. *Hydrolysis* refers to the reaction of a substance with water or its ions.

A common kind of hydrolysis reaction is the combination of the anion of a weak acid (from a salt) with H_3O^+ ions from water to form nonionized weak acid molecules.

This reaction of the anion of a weak monoprotic acid with water is commonly represented as

Recall that at 25°C,

in neutral solutions: $[H_3O^+] = [OH^-] = 1.0 \times 10^{-7}\ M$

in basic solutions: $[H_3O^+] < [OH^-] > 1.0 \times 10^{-7}\ M$

in acidic solutions: $[OH^-] < [H_3O^+] > 1.0 \times 10^{-7}\ M$

In Brønsted-Lowry terminology, anions of strong acids are very weak bases, whereas anions of weak acids are stronger bases. Conversely, cations of strong bases are very weak bases, whereas cations of weak bases are stronger acids. *The conjugate base of a strong acid is a very weak base - the conjugate base of a weak acid is a stronger base.*

Hydrochloric acid, a typical strong acid, is essentially completely ionized in dilute aqueous solutions.

Its conjugate base, the Cl^- ion, is a very weak base. It shows almost no tendency to combine with H_3O^+ ion to form nonionized HCl in dilute aqueous solutions.
The same is true for other strong acids and their anions.

By contrast, hydrofluoric acid (HF) is only slightly ionized in dilute aqueous solutions. Its conjugate base, the F^- ion, is a much stronger base than the Cl^- ion. So F^- ions combine with H_3O^+ ions to form nonionized hydrofluoric acid molecules.

Let us now consider dilute aqueous solutions of salts that contain no free acid or base. We have classified acids and bases as strong and weak. Based on this classification, we can identify four kinds of salts.

1. Salts of strong bases and strong acids

2. Salts of strong bases and weak acids

3. Salts of weak bases and strong acids

4. Salts of weak bases and weak acids

18-7 Salts of Strong Bases and Strong Acids

Aqueous solutions of *salts derived from strong acids and strong bases are neutral.* Potassium nitrate, KNO_3, is the salt of potassium hydroxide, KOH, and nitric acid, HNO_3.

Potassium nitrate solutions are neutral because neither ion reacts with water to upset the H_3O^+/OH^- balance.

18-8 Salts of Strong Bases and Weak Acids

Soluble *salts of strong bases and weak acids hydrolyze to produce basic solutions* because anions of weak acids react with water to form hydroxide ions. Sodium hypochlorite, $NaClO$, is the salt of sodium hydroxide, $NaOH$, and hypochlorous acid, $HClO$.

The above equations can be combined into a single equation that represents the reaction

The equilibrium constant for this reaction, called a *hydrolysis constant*, is written as

For the first time, we can calculate equilibrium constants using other known equilibrium constants. Hydrolysis constants can also be determined experimentally. The values obtained from experiment agree with calculated values.

If we multiply the above expression by $[H^+]/[H^+]$, an algebraic manipulation that doesn't change the value of the expression, we have

We recognize that this expression can also be expressed as

Now we can calculate the hydrolysis constant for the hypochlorite ion, ClO^-.

We can do the same for the anion of any weak monoprotic acid and find that

K_a is the ionization constant for the weak monoprotic acid from which the anion is derived. K_b, is the hydrolysis constant for the anion.

The relationship, $\mathbf{K_w = K_a K_b}$ is valid for all conjugate acid-base pairs in aqueous solutions.

If we know the value for either K_a *or* K_b *for any conjugate acid-base pair*, the other can be calculated.

306

Example 18-16:   Calculate hydrolysis constants for the following anions of weak acids.

a.   F⁻, fluoride ion, the anion of hydrofluoric acid, HF. For HF, $K_a = 7.2 \times 10^{-4}$.

b.   CN⁻, cyanide ion, the anion of hydrocyanic acid, HCN. For HCN,
     $K_a = 4.0 \times 10^{-10}$.

These calculations tell us that cyanide ions, CN⁻, hydrolyze to a much greater extent than fluoride ions, F⁻, because HCN is a weaker acid than HF.

Example 18-17:   Calculate [OH⁻], pH, and percent hydrolysis for the hypochlorite ion in 0.10 M sodium hypochlorite, NaClO, solution. "Clorox", "Purex", etc., are 5%, sodium hypochlorite solutions.

The equation for the hydrolysis of the hypochlorite ion, its hydrolysis constant expression, and the algebraic representations of the equilibrium concentrations of the various species follow (K_b for ClO⁻ = 2.9×10^{-7}). We let x M = mol/L of ClO⁻ that hydrolyze.

$$ClO^- + H_2O \rightleftharpoons HClO^- + OH^-$$

Initial []'s

Change _____

At equil: []'s

Substitution into the hydrolysis constant expression gives

$$K_b = \frac{[HClO][OH^-]}{[ClO^-]} = 2.9 \times 10^{-7}$$

The percent hydrolysis for the hypochlorite ion may be represented as

You should do a similar calculation for 0.10 M NaF solution and the compare the results from 0.10 M sodium fluoride and 0.10 M sodium hypochlorite solutions.

Solution	K_a (parent acid)	K_b (for anion)	[OH⁻]	pH	% hydrolysis
0.10 M NaF	_____	_____	____	__	_____
0.10 M NaClO	_____	_____	____	__	_____

The 0.10 M NaClO solution is more basic than 0.10 M NaF solution because ClO⁻ is a stronger base than F⁻, i.e., HClO is a weaker acid than HF.

18-9   Salts of Weak Bases and Strong Acids

Aqueous solutions of salts of weak bases and strong acids are acidic because cations derived from weak bases react with water to produce hydrogen ions. Consider aqueous solutions of ammonium bromide, NH_4Br, the salt of aqueous ammonia, NH_3, and hydrobromic acid, HBr.

The reaction may be represented more simply as

In even more simplified form the reaction is often represented as

The hydrolysis constant expression is

Multiplication of the hydrolysis constant expression by $[OH^-]/[OH^-]$ gives

which we recognize as

In the simplest possible form, the equation for the hydrolysis of the ammonium ion, NH_4^+, and its hydrolysis constant expression are represented as

Example 18-18:   Calculate $[H^+]$, pH, and percent hydrolysis for the ammonium ion in 0.10 M ammonium bromide, NH_4Br, solution.

The equation for the hydrolysis of the ammonium ion and the representation of equilibrium concentrations follow. We let x M = mol/L of NH_4^+ that hydrolyze.

$$NH_4^+ \quad + \quad H_2O \quad \rightleftharpoons \quad NH_3 \quad + \quad H_3O^+$$

Initial []'s

Change

Equilibrium []'s

Substitution into the hydrolysis constant expression gives

$$K_a = \frac{[NH_3][H_3O^+]}{[NH_4^+]} = 5.6 \times 10^{-10}$$

The percent hydrolysis of the ammonium ion in 0.10 M NH$_4$Br solution is

Identical results are obtained for all 0.10 M solutions that contain *only salts of aqueous ammonia and strong monoprotic acids.*

18-10   Salts of Weak Bases and Weak Acids

The fourth class of salts is soluble salts of weak bases and weak acids. Anions of weak acids hydrolyze to give basic solutions. Cations of weak bases hydrolyze to give acidic solutions. Thus, salts of weak bases *and* weak acids contain cations and anions that hydrolyze. Whether the resulting solutions are neutral, basic, or acidic depends on the values of the hydrolysis constants for the cation and anion of the particular salt. We divide the salts of weak bases and weak acids into three classes based upon relative values for the ionization constants for their parent weak acids and weak bases.

Salts of Weak Bases and Weak Acids for which K$_b$ = K$_a$

The common example of this type of salt is ammonium acetate, NH$_4$CH$_3$COO, the salt of aqueous ammonia, NH$_3$, and acetic acid, CH$_3$COOH. The ionization constants for aqueous ammonia and acetic acid are both 1.8×10^{-5}. The ammonium ion hydrolyzes to produce H$^+$ ions. Its hydrolysis constant is

The acetate ion hydrolyzes to produce OH$^-$ ions. Its hydrolysis constant is

From the fact that the hydrolysis constants for ammonium ion and acetate ion are equal, we predict that aqueous solutions of ammonium acetate are neutral. This is because hydrolysis of ammonium ions produces exactly as many H$^+$ ions as the hydrolysis of acetate ions produces OH$^-$ ions.

Neutral aqueous solutions are produced by all salts for which the ionization constant for the parent weak acid is exactly equal to the ionization constant of the parent weak base. The condition that $K_b = K_a$ is highly restrictive, and there are very few examples.

Salts of Weak Bases and Weak Acids for which $K_b > K_a$

Solutions of all salts of weak bases and weak acids for which K_b is greater than K_a are basic. This is because the anion of the weaker acid hydrolyzes to a greater extent than the cation of the stronger base.

Consider ammonium hypochlorite, NH_4ClO, the salt of aqueous ammonia and hypochlorous acid. The ionization constant for the parent weak base is greater than the ionization constant for the parent weak acid. K_b for $NH_3 = 1.8 \times 10^{-5}$, K_a for $HClO = 3.5 \times 10^{-8}$. Both the cation and anion of the salt hydrolyze.

The ammonium ion hydrolyzes to produce H^+. Its hydrolysis constant is

The hypochlorite ion hydrolyzes to produce OH^-. Its hydrolysis constant is

Hydrolysis of ClO^- ions produces OH^- ions to a greater extent than hydrolysis of NH_4^+ ions produces H^+ ions. Therefore, the solution is basic, as are *all* solutions derived from acids and bases for which parent $K_b > K_a$.

Salts of Weak Bases and Weak Acids for $K_b < K_a$

Salts of weak bases and weak acids for which K_a is greater than K_b are acidic because the cation of the weaker base hydrolyzes to a greater extent than the anion of the stronger acid.

Consider a solution of trimethylammonium fluoride, $(CH_3)_3NHF$. K_a for $HF = 7.2 \times 10^{-4}$ and K_b for trimethylamine $(CH_3)_3N = 7.4 \times 10^{-5}$. Both the cation, $(CH_3)_3NH^+$, and the anion, F^-, hydrolyze.

The trimethylammonium ion hydrolyzes to produce H^+ ions. Its hydrolysis constant is

The fluoride ion hydrolyzes to produce OH^- ions. Its hydrolysis constant is

Hydrolysis of trimethylammonium ions, $(CH_3)_3NH^+$, produces H^+ ions to a greater extent than hydrolysis of F^- ions produces OH^- ions. Therefore the solution is acidic, as are *all* such solutions.

Let us summarize our discussion of hydrolysis reactions. We have studied only two kinds of reactions, namely

    (1)   reactions of anions of weak monoprotic acids (from a salt) with water to form free molecular acids and OH^-.

    (2)   reactions of cations of weak monoprotic bases (from a salt) with water to form free molecular bases and H^+.

Aqueous solutions of *salts of strong acids and strong soluble bases are neutral* because neither anions of strong acids nor cations of strong bases hydrolyze.

Aqueous solutions of *salts of strong bases and weak acids are basic* because anions of weak acids hydrolyze to produce OH^-. Cations of strong soluble bases do not hydrolyze.

Aqueous solutions of *salts of weak bases and strong acids are acidic* because cations of weak bases hydrolyze to produce H^+. Anions of strong acids do not hydrolyze.

Aqueous solutions of *salts of weak acids and weak bases may be neutral, basic, or acidic*, depending upon the relative strengths of their parent weak acids and weak bases.

18-11 Salts that Contain Small, Highly-Charged Cations

Cations of soluble weak bases hydrolyze to produce acidic solutions. So do small highly-charged cations, i.e., *cations related to the insoluble bases (metal hydroxides)*. As described in Chapter 14, all ions are hydrated in aqueous solutions. Consider a solution of $Be(NO_3)_2$. It is thought that Be^{2+} ions are tetrahydrated and sp^3 hybridized.

Four water molecules are bound to each Be^{2+} ion by coordinate covalent bonds. This results from the donation of an unshared pair of electrons on each oxygen atom into a vacant sp^3 hybrid orbital on Be^{2+}. The positively-charged Be^{2+} ions draw the electrons in the O-H bonds closer to the oxygen atoms than they (electrons) are in uncoordinated (free) water molecules. This weakens the O-H bonds and allows them to be broken to form (hydrated) H^+ and OH^- ions more readily than in uncoordinated water molecules. We represent the hydrolysis of $[Be(OH_2)_4]^{2+}$ ions as

In condensed form it is represented as

or, even more simply as

We understand that the Be^{2+} ion is tetrahydrated and that $[Be(OH)]^+$ has three water molecules coordinated to it.

Similar hydrolysis reactions occur in all solutions of salts derived from insoluble metal hydroxides. Generally, the extent of hydrolysis increases as the charge density (charge-to-size ratio) of the cation increases.

The hydrolysis constant expression for $[Be(OH_2)_4]^{2+}$ and its value are

or, more simply

Table 18-10 lists several small highly-charged (hydrated) metal ions, their (simple) ionic radii, and their hydrolysis constants, K_a.

Example 18-19: Calculate the pH and percent hydrolysis in 0.10 M aqueous $Be(NO_3)_2$ solution.

The equation for the hydrolysis reaction and the representations of concentrations of the various species follow. We let x M = mol/L of Be^{2+} ions that hydrolyze.

Substituting these concentrations into the hydrolysis constant expression gives

Now we calculate the percent hydrolysis of Be^{2+}.

314

We see that 0.10 M Be(NO$_3$)$_2$ solution is nearly as acidic 0.10 M CH$_3$COOH solution (Example 18-11, text).

	[H$_3$O$^+$]	pH	% hydrolysis or % ionization	
0.10 M Be(NO$_3$)$_2$	_____	_____	_____	% hydrolyzed
0.10 M CH$_3$COOH	1.3×10^{-3} M	2.89	1.3	% ionized

Synthesis Question

Rain water is slightly acidic because it absorbs carbon dioxide from the atmosphere as it falls from clouds. (Acid rain is even more acidic because it absorbs acidic anhydride pollutants such as NO$_2$ and SO$_3$ as it falls to earth.) If the pH of a stream is 6.5 and all of the acidity comes from CO$_2$, how many CO$_2$ molecules did a spherical drop of rain having a diameter of 6.0 mm absorb in its fall to earth?

Group Question

A common food preservative in citrus flavored drinks is sodium benzoate, the sodium salt of benzoic acid. How does this chemical compound behave in solution so that it preserves the flavor of citrus drinks?

Chapter Nineteen

IONIC EQUILIBRIA II: BUFFERS AND TITRATION CURVES

19-1   The Common Ion Effect and Buffer Solutions

The term *common ion effect* describes solutions in which the same ion is produced by two different compounds. Buffer solutions resist changes in pH as acids or bases are added. Buffer solutions are special cases of the common ion effect. Because buffer solutions resist changes in pH, they must be able to react with both acids and bases. The two most common kinds of buffer solutions are

1.   Solutions of a weak acid plus a soluble ionic salt of the weak acid.

2.   Solutions of a weak base plus a soluble ionic salt of the weak base.

Weak Acids Plus Salts of Weak Acids

Consider a solution that contains acetic acid, CH_3COOH, and sodium acetate, $NaCH_3COO$, a soluble ionic salt of acetic acid.

In solutions containing CH_3COOH and $NaCH_3COO$ in *reasonable* concentrations, both compounds serve as sources of acetate ions. The high concentration of acetate ions, from sodium acetate, shifts the equilibrium for the ionization of acetic acid far to the left.

318

Example 19-1:   Calculate the concentration of H^+ and the pH of a solution that is 0.15 M in acetic acid and 0.15 M in sodium acetate.

The appropriate equations and representations of the equilibrium concentrations are

The ionization constant expression for acetic acid is valid for *all* acetic acid solutions. Substituting these concentrations into the K_a expression gives

$$K_a = \frac{[H^+][CH_3COO^-]}{[CH_3COOH]} = 1.8 \times 10^{-5}$$

Apply the basic simplifying assumption ideas

Making these assumptions gives

To see just how much the acidity of 0.15 M CH_3COOH solution is reduced by the addition of 0.15 mole of $NaCH_3COO$ per liter, refer back to Example 18-11.

Solution	$[H^+]$	pH
0.15 M CH_3COOH	_____	_____
0.15 M CH_3COOH and 0.15 M $NaCH_3COO$	_____	_____

We see that the H^+ concentration is _____ times greater in the 0.15 M CH_3COOH solution than in the 0.15 M CH_3COOH/0.15 M $NaCH_3COO$ buffer solution.

319

We can show the general relationship between the acidity of a solution containing a weak acid plus a salt of the weak acid and the ionization constant for the weak acid. We write the equation for the ionization of a weak *monoprotic* acid in general terms

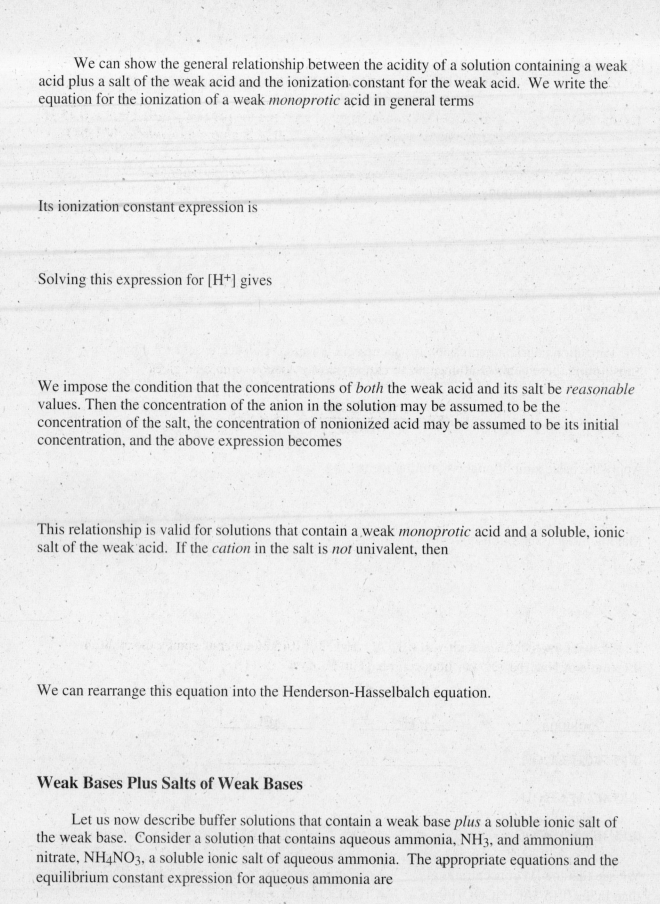

Its ionization constant expression is

Solving this expression for [H$^+$] gives

We impose the condition that the concentrations of *both* the weak acid and its salt be *reasonable* values. Then the concentration of the anion in the solution may be assumed to be the concentration of the salt, the concentration of nonionized acid may be assumed to be its initial concentration, and the above expression becomes

This relationship is valid for solutions that contain a weak *monoprotic* acid and a soluble, ionic salt of the weak acid. If the *cation* in the salt is *not* univalent, then

We can rearrange this equation into the Henderson-Hasselbalch equation.

Weak Bases Plus Salts of Weak Bases

Let us now describe buffer solutions that contain a weak base *plus* a soluble ionic salt of the weak base. Consider a solution that contains aqueous ammonia, NH$_3$, and ammonium nitrate, NH$_4$NO$_3$, a soluble ionic salt of aqueous ammonia. The appropriate equations and the equilibrium constant expression for aqueous ammonia are

320

We see that the high concentration of NH_4^+ ions, from NH_4NO_3 shifts the equilibrium for the ionization of aqueous ammonia far to the left.

Example 19-2:   Calculate the concentration of OH^- and the pH of a solution that is 0.15 M in aqueous ammonia, NH_3, *and* 0.30 M in ammonium nitrate, NH_4NO_3.

The appropriate equations and algebraic representations of the concentrations are

Substitution into the ionization constant expression for aqueous ammonia gives

$$K_b = \frac{[NH_4^+][OH^-]}{[NH_3]} = 1.8 \times 10^{-5}$$

In Example 18-13, we calculated $[OH^-]$ and the pH of 0.15 M aqueous ammonia solution. Comparison with the results obtained here is instructive.

Solution	[OH⁻]	pH
0.15 M aq NH_3	_____	_____
0.15 M aq NH_3 and 0.30 M NH_4NO_3	_____	_____

The concentration of OH⁻ in 0.15 M aqueous NH$_3$ is _____ times greater than in the buffer solution that is 0.15 M in NH$_3$ *and* 0.30 M in NH$_4$NO$_3$.

We may derive a general relationship for buffer solutions that contain a weak base plus a salt of the weak base. We write the equation for the ionization of a weak base in general terms.

The ionization constant expression is

Solving this expression for [OH⁻] gives

So long as we consider salts that contain univalent anions,

For salts that contain divalent or trivalent anions, we have

We can rearrange this equation into the Henderson-Hasselbalch equation.

19-2   Buffering Action

Buffer solutions resist changes in pH. The more acidic component of a buffer solution reacts with added base. The more basic component reacts with added acid.

Example 19-3:   If 0.020 mole of gaseous HCl is added to 1.00 liter of a buffer solution that is 0.100 M in aqueous ammonia and 0.200 M in ammonium chloride, how much does the pH change? Assume no volume change due to the addition of gaseous HCl.

First we calculate the pH of the original buffer solution.

Now we calculate the concentrations of all species after the addition of HCl. The HCl reacts with some of the NH_3.

$$HCl \quad + \quad NH_3 \quad \rightarrow \quad NH_4Cl$$

Initial []'s

<u>Change due to reaction</u>

After reaction []'s

We know the final concentrations of NH_3 and NH_4Cl so we can calculate [OH^-] easily.

Finally, we calculate the *change* in pH.

Example 19-4:   If 0.020 mole of solid NaOH is added to 1.00 liter of a buffer solution that is 0.100 M in aqueous ammonia and 0.200 M in ammonium chloride, how much does the pH change? Assume no volume change due to the addition of solid NaOH.

As in Example 19-3, the pH of the original buffer solution is _____. We calculate the concentrations of all species after addition of NaOH. NaOH reacts with NH_4^+ ions.

$$NH_4Cl \ + \ NaOH \ \rightarrow \ NH_3 \ + \ H_2O \ + \ NaCl$$

Initial []'s

Change due to reaction
_____

After reaction []'s

We know the concentrations of NH_3 and NH_4Cl so we calculate [OH⁻] and/or pOH.

Finally, we calculate the change in pH.

Let us summarize the results of Examples 19-3 and 19-4. See Table 19-3.

1.0 L of buffer solution that is 0.100 M aq NH_3 and 0.200 M NH_4Cl

pH

add 0.020 mol NaOH $\rightarrow$ _____ ΔpH = _____

_____

add 0.020 mol HCl $\rightarrow$ _____ ΔpH = _____

Notice that the pH changes only slightly in each case.

19-3   Preparation of Buffer Solutions

Buffer solutions may be prepared by mixing other solutions. When solutions are mixed, the concentrations of the dissolved species change because the volume in which they are contained changes.

Example 19-5:   Calculate the concentration of H^+ and the pH of a solution prepared by mixing 200. mL of 0.150 M acetic acid and 100 mL of 0.100. M sodium hydroxide solutions.

The amounts of acetic acid and sodium hydroxide (*before any reaction occurs*) in the solution are

Sodium hydroxide and acetic acid react in a 1:1 mole ratio.

$$NaOH \ + \ CH_3COOH \ \rightarrow \ NaCH_3COO \ + \ H_2O$$

Initial []'s

Change due to reaction _____

After reaction []'s

After mixing, the concentrations of CH_3COOH and $NaCH_3COO$ are (*in 300 mL of solution*)

Substitution into the ionization constant expression (or the Henderson-Hasselbalch equation) for CH_3COOH gives

On occasion it is desirable to prepare a buffer solution of a given pH.

Example 19-6:   Calculate the number of moles of solid ammonium chloride, NH_4Cl, that must be used to prepare 1.00 liter of a buffer solution that is 0.10 M in aqueous ammonia, and that has a pH of 9.15.

Because pH = 9.15 and pOH = 4.85,

The appropriate equations and the representation of equilibrium concentrations are

Substitution into the ionization constant expression (or the Henderson-Hasselbalch equation) for aqueous ammonia gives

$$K_b = \frac{[NH_4^+][OH^-]}{[NH_3]} = 1.8 \times 10^{-5}$$

326

19-4 Acid-Base Indicators

The *equivalence point* in an acid-base titration is the point at which chemically equivalent amounts of acid and base have reacted (Section 11-2). We can use the color change of an indicator to signal the *end point* of the titration. Ideally, the two should coincide. Visual indicators are intensely colored substances. Different indicators change color over different pH ranges. Many are very complex weak organic acids or bases. If we use an indicator that is a weak acid as an example, we may represent such an indicator as

$$HIn \;\rightleftharpoons\; H^+ + In^-$$

In^- = complex organic
part of molecule

The equilibrium constant expression for such an indicator may be represented as

The value of the ionization constant for an indicator, K_{HIn}, determines the pH range over which its color change occurs. Rearranging the above expression gives

We see that the color of such an indicator depends upon (1) $[H^+]$, and (2) the value of K_a for the indicator.

As a rule-of-thumb, when the ratio $[In^-]/[HIn] \geq 10$, color 2 is observed. Conversely, when the ratio $[In^-]/[HIn] \leq 1/10$, color 1 is observed. Because $[H^+]$ must change by a factor of about 100 to go from one of these ratios to the other, the pH range over which a color change occurs is about (log 100 =) 2 pH units. The specific pH range depends upon K_a (or K_b) for the indicator. The following table lists four common indicators. See Figures 19-1, 19-2, and Table 19-4.

Some Acid-Base Indicators

Indicator	Color in More Acidic Range	pH Range	Color in More Basic Range
Methyl violet (MV)	yellow	0-2	purple
Methyl orange (MO)	pink	3.1-4.4	yellow
Litmus	red	4.7-8.2	blue
Phenolphthalein (PP)	colorless	8.3-10.00	red

TITRATION CURVES

19-5   Strong Acid/Strong Base Titration Curves

A plot of pH *vs* amount (or volume) of acid or base added in an acid-base titration displays graphically the change in pH as acid or base is added to a solution. It shows clearly how pH changes near the *equivalence point*, the point at which equivalent amounts of acid and base have reacted. An indicator is chosen so that its color change occurs at, or very near, the equivalence point. (See Table 19-4.) Plots of pH vs amount of acid or base added are called *titration curves*. Reactions of acids with bases are *neutralization* reactions.

Consider the titration of 100.0 mL of 0.100 *M* perchloric acid, $HClO_4$, solution with 0.100 *M* potassium hydroxide, KOH, solution. Let us calculate and plot the pH of the solution vs mL of 0.100 *M* KOH solution added. The reaction ratio is a 1:1 mole ratio.

Before any KOH solution is added, the pH of the 0.100 *M* $HClO_4$ solution is _____.

When 20.0 mL of 0.100 *M* KOH solution has been added, the pH is _____.

$$HClO_4 \;+\; KOH \;\rightarrow\; KClO_4 \;+\; H_2O$$

Starting []'s

Change
_____

After reaction []'s

When 50.0 mL total of 0.100 *M* KOH solution has been added (mid-point of titration) the pH is _____.

$$HClO_4 \;+\; KOH \;\rightarrow\; KClO_4 \;+\; H_2O$$

Starting []'s

Change
_____

After reaction []'s

When 90.0 mL total of 0.100 M KOH solution has been added, the pH is _____.

$$HClO_4 \; + \; KOH \; \rightarrow \; KClO_4 \; + \; H_2O$$

Starting []'s

Change

After reaction []'s

When 100.0 mL total of 0.100 M KOH solution has been added (*the equivalence point*) the pH is
_____.

$$HClO_4 \; + \; KOH \; \rightarrow \; KClO_4 \; + \; H_2O$$

Starting []'s

Change

After reaction []'s

When 101.0 mL total of 0.100 M KOH solution has been added the pH of the solution is
_____.

$$HClO_4 \; + \; KOH \; \rightarrow \; KClO_4 \; + \; H_2O$$

Starting []'s

Change

After reaction []'s

We have calculated only a few points on the titration curve. A plot of these and a few other points shows the shape of the titration curve clearly.

mL of 0.100 M KOH added	mmol base added	mmol excess acid or base		pH
0.0 mL	0	10.0	H+	1.00
20.0 mL	2.00	8.0		1.17
50.0 mL	5.00	5.0		1.48
90.0 mL	9.00	1.0		2.28
99.0 mL	9.90	0.10		3.30
99.5 mL	9.95	0.05		3.60
100.0 mL	10.00	0		7.00
100.5 mL	10.05	0.05	OH-	10.40
110.0 mL	11.00	1.0		11.68
120.0 mL	12.00	2.0		11.96

Titration Curve for $HClO_4$ versus KOH

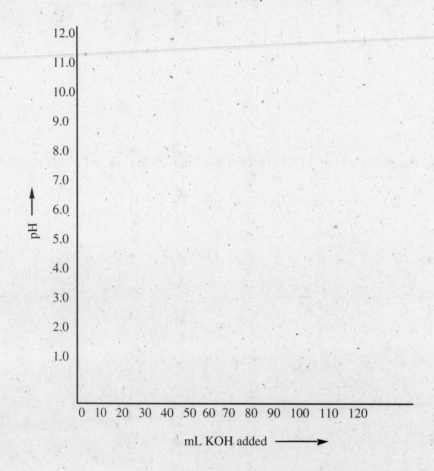

330

Note that pH changes rapidly as very small amounts of KOH are added just before *and* just after the equivalence point (pH = 7.00). Thus, indicators that change within the pH range of approximately 4-10 can be used for titrations of strong acids and bases. The data and curves of all strong acid/strong base titrations are identical to those involving HCl and NaOH (Table 19-5 and Figure 19-3a). Titrations of strong acids and strong bases are the simplest kind.

The titration curve for a strong acid added to a solution of a strong base is very similar, but inverted. See Figure 19-3b.

19-6   Weak Acid/Strong Base Titration Curves

Consider the titration of 100.0 mL of 0.100 M acetic acid, CH_3COOH, solution with 0.100 M KOH solution (remember the strong solution is *always added* to the weak solution). The acid and base react in a 1:1 mole ratio.

However, there are two complicating factors in this titration.

The first is due to the fact that before the equivalence point, both CH_3COOH and KCH_3COO are present (a buffer solution).

Before any KOH solution is added, the pH of the 0.100 M CH_3COOH solution is

After 20.0 mL of KOH solution has been added, the pH is _____.

$$KOH \ + \ CH_3COOH \ \rightarrow \ KCH_3COO \ + \ H_2O$$

Initial []'s

Change due to reaction

After reaction []'s

Other points before the equivalence point are calculated like this one.

The second complication arises at the equivalence point. The solution is now 0.0500 M in KCH_3COO, the salt of a strong base and a weak acid that hydrolyzes to give a basic solution. Therefore, pH $\neq$ 7.00 at the equivalence point. (Refer to Example 19-2 in this outline and Example 19-1 in the text for similar calculations.)

$$KOH \quad + \quad CH_3COOH \quad \rightarrow \quad KCH_3COO \quad + \quad H_2O$$

Initial []'s

Change due to reaction _____

After reaction []'s

Beyond the equivalence point the basicity of the solution is determined by the excess KOH, just as it was in the titration of a strong acid with a strong soluble base.

After 110.0 mL of KOH has been added, the pH is _____.

$$KOH \quad + \quad CH_3COOH \quad \rightarrow \quad KCH_3COO \quad + \quad H_2O$$

Initial []'s

Change due to reaction _____

After reaction []'s

From these points, and a few that are supplied, we can plot the titration curve.

mL of 0.100 M KOH added	mmol base added	mmol excess acid or base	pH
0.0 mL	0	10.0 CH$_3$COOH	2.89
20.0 mL	2.00	8.00	4.15
50.0 mL	5.00	5.00	4.74
75.0 mL	7.50	2.50	5.22
90.0 mL	9.00	1.00	5.70
95.0 mL	9.50	0.50	6.02
99.0 mL	9.90	0.10	6.74
100.0 mL	10.0	0	8.72
101.0 mL	10.1	1.0 OH$^-$	10.70
110.0 mL	11.0	1.0	11.68
120.0 mL	12.0	2.0	11.96

Titration Curve for CH$_3$COOH versus KOH

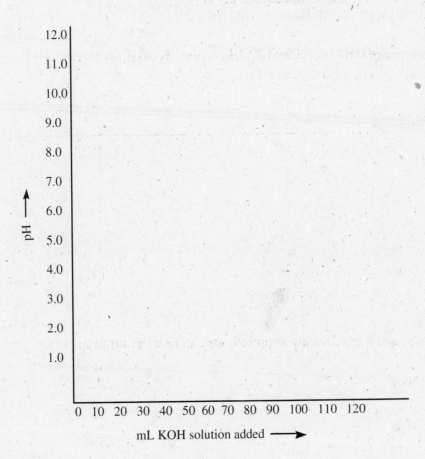

333

Note that pH = 8.72 at the equivalence point in this titration. Therefore, an indicator whose color change range spans pH = 8.72 is required. The vertical portion of this titration curve is shorter than in the case of $HClO_4$ vs KOH (strong acid/strong base), and so the choice of indictors is thereby restricted.

Titration curves for strong acids and weak bases are similar to those for strong bases and weak acids, but they are inverted. See Figure 19-5.

19-7   Weak Acid/Weak Base Titration Curves

Weak acid and weak base titration curves have such short vertical sections that they are not noticeable. See Figure 19-6. This is because the solution is buffered before *and* after the equivalence point. The equivalence point in such titrations cannot be detected with visual or color indicators. Other than the initial pH and the estimation of the pH at the equivalence point, pH calculations for such systems are beyond the scope of these texts.

19-8   Summary of Acid-Base Calculations

The summary of the acidic or alkaline solutions discussed in this and previous chapters are shown in Table 19-7. Review this table.

Synthesis Question

Bufferin is a commercially prepared medicine that is literally buffered aspirin. How could you buffer aspirin? Hint - what is aspirin?

Group Question

Blood is slightly basic, having a pH of 7.35 to 7.45. What chemical species causes our blood to be basic? How does our body regulate the pH of our blood?

335

Chapter Twenty

IONIC EQUILIBRIA III:
THE SOLUBILITY PRODUCT PRINCIPLE

In our discussion of ionic equilibria to this point, we have discussed compounds that are readily soluble in water. However, many compounds are only slightly soluble in water. These are often referred to as "insoluble" compounds. Nearly all compounds dissolve in water to at least some extent. As a rough rule-of-thumb, a compound that is soluble to the extent of 0.020 mole per liter, or more, is classified as soluble.

20-1   Solubility Product Constants

Suppose some solid silver chloride, AgCl, is placed in water and stirred vigorously. Careful experiments show that a small amount of silver chloride dissolves in water, and the silver chloride that dissolves is dissociated completely into its ions.

The equilibrium constant expression for this reaction is called a *solubility product constant, K_{sp}.* The activity of solid AgCl is 1. See Section 17-2.

In general terms, *the solubility product constant for a compound is the product of the concentrations of the constituent ions, each raised to the power that corresponds to the number of ions in one formula unit of the compound.* This is a statement of the solubility product principle.

338

Solubility data and solubility product constants are usually reported at 25°C. Consider the dissolution of the slightly soluble compound silver sulfide, Ag_2S, in water.

Its solubility product expression is

The dissolution of solid calcium phosphate, $Ca_3(PO_4)_2$, in water is represented as

Its solubility product constant expression is

Generally, the dissolution of a slightly soluble compound and its solubility product expression may be represented as

It there are more than two kinds of ions in the formula for a compound, the generalization applies. For example, the dissolution of slightly soluble calcium ammonium phosphate, $CaNH_4PO_4$, in water and its solubility product expression are represented as

Refer to Appendix H in the text for solubility product constants as needed.

20-2   Determination of Solubility Product Constants

Solubility product constants may be determined in a number of ways. For example, careful measurements show that 1.00-liter of saturated AgCl solution contains 0.00192 g of *dissolved* AgCl. From this information we can calculate the solubility product for AgCl.

Example 20-1:   One liter of saturated silver chloride solution contains 0.00192 g of dissolved AgCl at 25°C. Calculate the molar solubility of, and K_{sp} for, AgCl.

The *molar solubility* of silver chloride can be calculated from the data given.

The equation for the dissociation of silver chloride and its solubility product expression are

Substitution into the solubility product expression gives

This solubility product expression and constant are applicable to *all saturated silver chloride solutions at 25°C*. The origin of the silver and chloride ions is of no importance.

Example 20-2:   One liter of saturated calcium fluoride solution contains 0.0167 gram of CaF_2 at 25°C. Calculate the molar solubility of, and K_{sp} for, CaF_2.

First, we calculate the molar solubility of CaF_2.

340

We know the molar solubility of CaF_2 in water, and so we can represent the concentrations of ions in saturated CaF_2 solution. This lets us calculate K_{sp}.

The molar solubility of calcium fluoride is 2.14×10^{-4} mol/liter and its solubility product is 3.9×10^{-11}. Let us compare the values obtained in Examples 20-1 and 20-2.

Compound	Molar Solubility	Solubility Product
AgCl	1.34×10^{-5} M	1.8×10^{-10}
CaF$_2$	2.14×10^{-4} M	3.9×10^{-11}

We see that the *molar solubility* of CaF_2 is greater than the molar solubility of AgCl, but that K_{sp} for AgCl is greater than K_{sp} for CaF_2. The solubility product expression for CaF_2 contains a squared term, i.e., $[F^-]^2$. We must always exercise care in comparing solubility products of compounds!

20-3 Uses of Solubility Product Constants

If the solubility product of a compound is known, the solubility of the compound in water at 25°C can be calculated conveniently.

Example 20-3: Calculate the molar solubility of barium sulfate, $BaSO_4$, in pure water and the concentrations of barium and sulfate ions in saturated barium sulfate at 25°C. $K_{sp} = 1.1 \times 10^{-10}$

The equation for the dissociation of barium sulfate in water shows that each formula unit of barium sulfate that dissolves produces one Ba^{2+} and one SO_4^{2-} ion. We let x mol/L represent the molar solubility of $BaSO_4$ (as well as the concentrations of its ions).

Substitution into the solubility product expression, and solving for x, gives the molar solubility of $BaSO_4$ and the concentrations of its ions in saturated solution at 25°C.

The mass of $BaSO_4$ in 1.00 L of saturated solution can now be calculated.

Example 20-4:   The solubility product for magnesium hydroxide, $Mg(OH)_2$, is 1.5×10^{-11}. Calculate the molar solubility of magnesium hydroxide and the pH of saturated magnesium hydroxide solution at 25°C.

The equation for the dissociation of magnesium hydroxide, the representation of its molar solubility, and the concentrations of its ions follow. We let $x\ M$ = molar solubility of $Mg(OH)_2$.

Substitution into the solubility product expression gives

The Common Ion Effect in Solubility Calculations

The *common ion effect* is applicable to solubility equilibria, just as it is to other ionic equilibria.

Example 20-5:   Calculate the molar solubility of barium sulfate, $BaSO_4$, in 0.010 M sodium sulfate, Na_2SO_4, solution at 25°C. Compare this to the solubility of $BaSO_4$ in pure water (Example 20-3).

First, we write equations for the processes that occur and represent the concentrations of ions. We let x mol/L represent the molar solubility of $BaSO_4$ in *this solution*.

Now we substitute the concentrations into the K_{sp} expression and solve for x.

The molar solubility of $BaSO_4$ in 0.010 M Na_2SO_4 solution is 1.1×10^{-8} M. The molar solubility of $BaSO_4$ in pure water, 1.0×10^{-5} M, is about 900 times *greater*.

The Reaction Quotient in Precipitation Reactions

One important use of solubility product constants is in the calculation of the concentrations of ions that can exist in solutions, and whether or not a precipitate will form in a given solution.

Example 20-6:   We mix 100. mL of 0.010 M potassium sulfate, K_2SO_4, and 100. mL of 0.10 M lead(II) nitrate, $Pb(NO_3)_2$, solutions. Will a precipitate form?

First, we consider the kinds of compounds mixed to determine *if* a reaction *can* occur. Both K_2SO_4 and $Pb(NO_3)_2$ are soluble ionic compounds. At the instant of mixing, the new solution contains four ions.

Having determined that $PbSO_4$ *may* precipitate, we shall calculate Q_{sp} for $PbSO_4$. The volumes of dilute aqueous solutions are additive. At the moment of mixing, the concentrations of Pb^{2+} and SO_4^{2-} ions are

We calculate Q_{sp} for $PbSO_4$

K_{sp} for $PbSO_4$ is 1.8×10^{-8}. We see that $Q_{sp} > K_{sp}$. Therefore solid $PbSO_4$ will precipitate until the ion concentrations just satisfy K_{sp} for $PbSO_4$.

For a precipitate to be visible, its K_{sp} must be exceeded by approximately 10^3. This is a rough rule-of-thumb. White, or very nearly colorless, compounds may not be visible until solubility products are exceeded by more than 10^3.

Frequently it is desirable to remove an ion from solution as nearly completely as practical by forming an insoluble compound. Solubility products can be used to calculate concentrations of ions remaining in solution after precipitation has occurred.

Example 20-7: Suppose we wish to remove mercury from an aqueous solution that contains a soluble mercury compound such as $Hg(NO_3)_2$. We can do this by precipitating mercury(II) ions in the insoluble compound HgS. What concentration of sulfide ions, from a soluble compound such as Na_2S, is required to reduce the Hg^{2+} concentration to 1.0×10^{-8} M? For HgS, $K_{sp} = 3.0 \times 10^{-53}$

The equation for the reaction of mercury(II) ions and sulfide ions, and the solubility product expression for mercury(II) sulfide are

To determine the concentration of S^{2-} required to reduce the concentration of Hg^{2+} to 1.0×10^{-8} M, we solve for $[S^{2-}]$ in the solubility product expression.

To reduce Hg^{2+} to 1.0×10^{-8} M (this is only 0.0000020 g Hg^{2+} per liter), we would add sufficient Na_2S to precipitate HgS until the *final* concentration of S^{2-} in this solution is 3.0×10^{-45} M. Because HgS is so very insoluble, only ever so slightly more than a stoichiometric amount of Na_2S is required to precipitate Hg^{2+} very nearly completely.

Example 20-8: Refer to Example 20-7. What volume of the solution (1.0×10^{-8} M Hg^{2+}) contains 1.0 g of mercury?

20-4 Fractional Precipitation

On occasion we wish to remove some ions from solution while leaving other ions with similar properties in the solution. The process is called *fractional precipitation*. Consider a solution that contains Cu^+, Ag^+, and Au^+ ions. Assume that the anions are NO_3^-. The dissolution equilibria and the K_{sp} expressions for their chlorides are

$$CuCl(s) \rightleftharpoons Cu^+ + Cl^- \qquad\qquad K_{sp} = [Cu^+][Cl^-] = 1.9 \times 10^{-7}$$

$$AgCl(s) \rightleftharpoons Ag^+ + Cl^- \qquad\qquad K_{sp} = [Ag^+][Cl^-] = 1.8 \times 10^{-10}$$

$$AuCl(s) \rightleftharpoons Au^+ + Cl^- \qquad\qquad K_{sp} = [Au^+][Cl^-] = 2.0 \times 10^{-13}$$

Example 20-9: If solid sodium chloride is slowly added to a solution that is 0.010 M each in Cu^+, Ag^+, and Au^+ ions, which compound precipitates first? Calculate the concentration of Cl^- required to initiate precipitation of each of these metal(I) chlorides.

The solubility product for AuCl is smallest, and so AuCl precipitates first.

We know the concentration of Au^+ in the solution. We can calculate the concentration of Cl^- required to initiate precipitation of AuCl.

Repeating this kind of calculation for silver chloride gives

345

For copper(I) chloride to precipitate

We have calculated the [Cl⁻] required

to precipitate AuCl, $[Cl^-] > 2.0 \times 10^{-11}\ M$

to precipitate AgCl, $[Cl^-] > 1.8 \times 10^{-8}\ M$

to precipitate CuCl, $[Cl^-] > 1.9 \times 10^{-5}\ M$

These calculations tell us that when sodium chloride is added slowly to a solution that is 0.010 M each in Cu^+, Ag^+, and Au^+ ions, AuCl precipitates first, AgCl second, and CuCl last.

We can calculate the amount of Au^+ precipitated before Ag^+ begins to precipitate, as well as the amounts of Au^+ and Ag^+ precipitated before Cu^+ begins to precipitate.

Example 20-10:   Calculate the percent of Au^+ ions that precipitate before AgCl begins to precipitate.

In Example 20-9, we found that [Cl⁻] > _____ M is necessary to begin precipitation of AgCl. This value of [Cl⁻] can be substituted into the solubility product expression for AuCl to determine [Au⁺] *remaining in solution* (i.e., unprecipitated) *just before AgCl begins to precipitate*.

The percent of Au^+ ions *unprecipitated* just before AgCl precipitates is

Therefore, _____% of the Au^+ ions precipitate *before* AgCl begins to precipitate.

Example 20-11:   Calculate the percent of Au^+ and Ag^+ ions that precipitate before CuCl begins
to precipitate.

The concentration of Cl^-, just before CuCl begins to precipitate, is _____ M (Example
20-9). We use this concentration of Cl^- to determine the concentrations of Au^+ and Ag^+ in the
solution at this point. The concentration of Au^+ *unprecipitated* is

The percent of Au^+ ions *unprecipitated* just *before* CuCl begins to precipitate is

Thus, _____ % of the Ag^+ ions precipitate before CuCl begins to precipitate.

A similar calculation for the concentration of Ag^+ ions unprecipitated *before* CuCl begins to
precipitate gives

The percent of Ag^+ ions *unprecipitated* just before CuCl begins to precipitate is

Thus, _____ % of the Ag^+ ions precipitate before CuCl begins to precipitate.

We have described the series of reactions that occurs when solid NaCl is added slowly to a
solution that is 0.010 molar each in Au^+, Ag^+, and Cu^+ ions. AuCl begins to precipitate first, and
99.89% of the Au^+ ions precipitate before any solid AgCl is formed. AgCl precipitates next;
99.99989% of the Au^+ ions and 99.905% of the Ag^+ ions precipitate before any solid CuCl
forms.

347

20-5   Simultaneous Equilibria Involving Slightly Soluble Compounds

   To this point we have considered most equilibria involving one equilibrium constant expression at a time. Let us now consider equilibria that involve two (or more) different equilibrium constant expressions simultaneously.

Example 20-12:   If 0.10 mole of ammonia and 0.010 mole of magnesium nitrate, $Mg(NO_3)_2$, are added to enough water to make one liter of solution, will magnesium hydroxide precipitate from the solution? For $Mg(OH)_2$, $K_{sp} = 1.5 \times 10^{-11}$; K_b for $NH_3 = 1.8 \times 10^{-5}$.

The question to be answered is - is the solubility product for magnesium hydroxide, $Mg(OH)_2$, exceeded in the solution? We calculate Q_{sp} for $Mg(OH)_2$ and compare it to K_{sp}.

$Mg(NO_3)_2$ is a soluble ionic compound, so we know that $[Mg^{2+}] = 0.010\ M$. Aqueous ammonia is a weak base that ionizes slightly. We can calculate $[OH^-]$.

We know the concentrations of both Mg^{2+} and OH^- ions, so we calculate Q_{sp} for $Mg(OH)_2$.

We see that $Q_{sp} > K_{sp} = 1.5 \times 10^{-11}$, and so $Mg(OH)_2$ precipitates.

   Most metal hydroxides are "insoluble" in water. Frequently, it is desirable to prepare basic solutions that contain fairly high concentrations of these metal ions. This is accomplished by buffering the solutions. This keeps $[OH^-]$ low enough that the K_{sp}'s of the metal hydroxides are not exceeded.

Example 20-13:   How many moles of solid ammonium chloride, NH$_4$Cl, must be used to prevent precipitation of Mg(OH)$_2$ in one liter of solution that is 0.10 M in aqueous ammonia *and* 0.010 M in magnesium nitrate, Mg(NO$_3$)$_2$? (Note the similarity between this problem and Example 20-12).

Because the precipitation of Mg(OH)$_2$ is to be prevented, we calculate the maximum [OH$^-$] that can exist in a solution that is also 0.010 M in Mg^{2+}. Substitution of [Mg^{2+}] into the K$_{sp}$ expression for Mg(OH)$_2$ allows us to calculate the maximum [OH$^-$].

Now we know the maximum [OH$^-$] that can exist in the solution, so we calculate the number of moles of NH$_4$Cl required to buffer 0.10 M aqueous ammonia so that [OH$^-$] does not exceed 3.9 × 10^{-5} M.

The appropriate equations, algebraic representations of equilibrium concentrations, and equilibrium constant expression are

We can check our values by calculating Q$_{sp}$ for Mg(OH)$_2$.

We see that: Q$_{sp}$ = 1.5 × 10^{-11} = K$_{sp}$ ∴ Equilibrium!

In Example 20-13 we used two equilibrium constants, the ionization constant for aqueous NH_3 and the solubility product for $Mg(OH)_2$. We can use a third equilibrium constant, the ion product for water, to calculate $[H^+]$ and the pH of the solution.

20-6 Dissolving Precipitates

Precipitates can be dissolved by reducing the concentrations of their ions so that their solubility product constants are no longer exceeded, i.e., so that $Q_{sp} < K_{sp}$. The precipitate dissolves until $Q_{sp} = K_{sp}$. This may be accomplished in several ways.

Converting an Ion to a Weak Electrolyte

1. Converting OH^- to H_2O. Insoluble metal hydroxides dissolve in acids. H^+ ions react with OH^- ions to form the weak electrolyte H_2O.

2. Converting NH_4^+ to NH_3. The NH_4^+ ions combine with OH^- ions in the saturated metal hydroxide solution. This will form the weak electrolytes NH_3 and H_2O.

3. Converting S_2^- to H_2S. Nonoxidizing acids dissolve most insoluble metal sulfides. The H^+ ion combine with S_2^- ions to form H_2S, a gas that bubbles out of the solution.

Converting an Ion to Another Species by a Redox Reaction

Very insoluble metal sulfides dissolve in hot nitric acid. Hot nitric acid oxidizes sulfide ions, S^{2-}, to elemental sulfur, S. For lead(II) sulfide,

Complex Ion Formation

The cations in many slightly soluble compounds form complex ions. In the photographic development process, exposed film is immersed in "hypo" (sodium thiosulfate, $Na_2S_2O_3$, solution). The "unactivated" silver chloride and silver bromide dissolve by forming soluble complex compounds. For AgBr, the reactions are

Another example is the dissolution of light blue copper(II) hydroxide, $Cu(OH)_2$, in excess aqueous NH_3 to form the dark blue complex ion, $[Cu(NH_3)_4]^{2+}$.

Synthesis Question

Most kidney stones are made of calcium oxalate, $Ca(C_2O_4)$. Patients who have their first kidney stones are given an extremely simple solution to stop further stone formation. They are told to drink six to eight glasses of water a day. How does this stop kidney stone formation?

Group Question

The cavities that we get in our teeth are a result of the dissolving of the enamel on our teeth, calcium hydroxy apatite. How does using a fluoride based toothpaste decrease the occurrence of cavities?

Chapter Twenty One

ELECTROCHEMISTRY

Electrochemistry deals with chemical changes produced by an electric current and with the production of electricity by chemical reactions. The study of electrochemistry has provided much of our knowledge about chemical reactions. The amount of electrical energy consumed or produced by electrochemical reactions can be measured accurately. Electrochemical reactions are *oxidation-reduction* reactions.

The two parts of a reaction are separated in electrochemical cells so that one-half of a reaction occurs at one electrode while the other half occurs at the other electrode.

There are two kinds of cells.

1. Electrical energy causes *nonspontaneous* chemical reactions to occur in **electrolytic cells.**

2. *Spontaneous* chemical reactions produce electrical energy in **voltaic cells.**

21-1   Electrical Conduction

Metals conduct an electric current well. This type of conduction of electricity is called *metallic conduction*. It involves the flow of electrons, with no similar movement of the atoms of the metal, and no obvious changes in the metal. Apparently, the electrons are being passed from one atom to the next.

In *ionic* or *electrolytic conduction* a current is conducted by the motion of ions, electrically charged particles, through a liquid. The positively charged ions, cations, move toward the negative electrode -- the negatively charged ions, anions, move toward the positive electrode. See Figure 21-1.

21-2   Electrodes

Let us establish the conventions. The *cathode* is the electrode at which *reduction* occurs, i.e., electrons are gained by some species. The *anode* is the electrode at which *oxidation* occurs, i.e., electrons are lost by some species. *Inert electrodes* do not react with the liquid or the products of an electrochemical reaction.

ELECTROLYTIC CELLS

Electrical energy causes *nonspontaneous* chemical reactions to occur in *electrolytic* cells. The process is called *electrolysis*. An electrolytic cell consists of a container for the reaction mixture and electrodes that are immersed in the reaction material. The electrodes are connected to a source of direct current.

21-3   The Electrolysis of Molten Sodium Chloride (the Downs Cell)
(See Figure 21-2)

The electrolysis of molten sodium chloride produces metallic sodium at one electrode and gaseous chlorine at the other. These observations allow us to diagram the cell and to deduce the electrode reactions as well as the overall reaction. The molten, silvery white metallic sodium remains liquid because its melting point is only 97.8°C. It floats because it is less dense than the molten NaCl.

The Down cell shown in Figure 21-2b is the main commercial method of producing metallic sodium. The sodium and chlorine are not allowed to come in contact with each other because they would react spontaneously to form sodium chloride. The electrode reactions and cell reaction for the electrolysis of molten sodium chloride are

Anode reaction

Cathode reaction
_____

Cell reaction

In *all electrolytic cells*, electrons are *forced* to flow from the positive electrode (anode) to the negative electrode (cathode).

21-4   The Electrolysis of Aqueous Sodium Chloride
(See Figure 21-3)

The electrolysis of an aqueous solution of NaCl (using *inert* electrodes) produces H_2 at one electrode and Cl_2 at the other; the solution becomes basic around the electrode at which H_2 is produced. Because chlorine is produced at one electrode, we conclude that this electrode (anode) reaction is as given below.

Because H_2 is produced at the other electrode, and the solution becomes basic around this electrode, we conclude that this electrode (cathode) reaction is as given below.

Balancing the electron transfer and adding the two electrode reactions give the cell reaction.

Anode reaction

Cathode reaction

Cell reaction

We can now diagram the cell and label the electrodes.

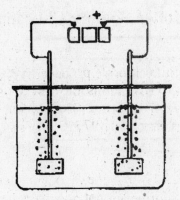

H_2 and OH^- ions are produced by the reduction of water. This tells us that H_2O is more easily reduced than Na^+ ions.

21-5   The Electrolysis of Aqueous Sodium Sulfate Solution
(See Figure 21-4)

Electrolysis of aqueous solutions of Na_2SO_4 (using *inert* electrodes) produces H_2 at one electrode, and the solution becomes basic around this electrode. This electrode (cathode) reaction is the same as before. O_2 is produced at the other electrode (anode), and the solution becomes acidic around this electrode.

Balancing the electron transfer and adding the electrode reactions gives the cell reaction.

Anode reaction

Cathode reaction _____

Cell reaction

We can diagram the cell and label the electrodes.

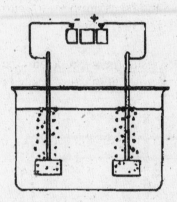

The electrolysis of aqueous Na_2SO_4 solution indicates that H_2O molecules are the species most easily reduced and also the species most easily oxidized in this solution. Neither potassium ions, Na^+, nor sulfate ions, SO_4^{2-}, is reduced or oxidized because other more easily oxidizable and reducible species (H_2O) are present.

We may generalize: **In all electrolytic cells the most easily reduced species is reduced and the most easily oxidized species is oxidized.**

21-6 Counting Electrons: Coulometry and Faraday's Law of Electrolysis

In 1832-33 Michael Faraday observed that *the amount of substance undergoing chemical reaction at each electrode during electrolysis is directly proportional to the amount of electricity that passes through the electrolytic cell*. This statement is known as *Faraday's Law of Electrolysis*. The term *faraday* refers to the amount of electricity that reduces one equivalent of a species at the cathode and oxidizes one equivalent of a species at the anode. This corresponds to the gain or loss of 6.022×10^{23} electrons. Thus *one faraday* corresponds to

A *coulomb* is defined as the amount of charge that passes a given point when a current of one ampere (A) flows for one second, i.e., 1 ampere = 1 coulomb/second (1 C/s). One faraday is also equivalent to

96,487 coulombs is often rounded off to 96,500 coulombs. We restate Faraday's Law. During electrolysis, one faraday of electricity (96,487 coulombs) reduces and oxidizes, respectively, one equivalent of the oxidizing agent and the reducing agent. This corresponds to the passage of 6.022×10^{23} electrons through an electrolytic cell.

Example 21-1:   Calculate the mass of palladium produced by the reduction of palladium(II) ions during the passage of 3.20 amperes of current through a solution of palladium(II) sulfate for 30.0 minutes.

The equation for the reduction of Pd^{2+} ions at the *cathode* and the equivalence between amount of electricity passed through the cell and amount of Pd produced are

Example 21-2:   Calculate the volume of oxygen (measured at STP) produced by the oxidation of water in Example 21-1.

The equation for the oxidation of H_2O and the equivalence between the amount of electricity passed through the cell and the amount of O_2 produced at the *anode* are

21-7   Commercial Applications of Electrolytic Cells

Some metals are obtained from their ores (by chemical reduction) in an impure form, and so further purification is usually necessary. Copper is one industrially important metal that is refined electrolytically. Impure copper contains some active metals as well as very small amounts of (very valuable) less active metals such as silver, gold, and platinum.

Very thin sheets of pure copper are made cathodes by connecting them to the negative terminal of a direct-current generator. Large bars of impure copper are made anodes. The supporting electrolyte is copper(II) sulfate and dilute H_2SO_4. The impure copper anode dissolves (is oxidized) to form Cu^{2+} ions, while Cu^{2+} ions are reduced to metallic copper at the (pure copper) cathode.

Anode reaction

Cathode reaction

Net reaction

Although *no net reaction occurs* in the electrolytic refining of copper, the result is that large pieces of impure copper are converted to small pieces of impure copper while small pieces of pure copper become large pieces of pure copper. (See Figure 21-5.)

Active metals are oxidized to cations that are more difficult to reduce than Cu^{2+}.

Electroplating with metals such as copper is similar to electrolytic refining. The object to be electroplated with a metal is made the cathode. For example, copper can be plated onto family mementos, chromium onto automobile bumpers, and silver or gold onto watch bands.

VOLTAIC OR GALVANIC CELLS

A *voltaic* or *galvanic cell* is an electrochemical cell in which a *spontaneous* oxidation-reduction reaction produces electrical energy. The two halves of the reaction are separated so that the electron transfer is forced to occur along a metallic surface and a potential difference is created. Common examples include

21-8   The Construction of Simple Voltaic Cells

A *half-cell* consists of the oxidized and reduced forms of an element (or other more complex species) in contact with each other. A common kind of a half-cell consists of a piece of metal immersed in a solution of its ions. Electrical contact between the two electrodes is frequently made by a wire and a *salt bridge*. A common kind of salt bridge is prepared by bending a piece of glass tubing twice at right angles. The tube is then inverted, filled with hot saturated potassium chloride/5% agar-agar solution, and allowed to cool. The cool mixture "sets" to about the consistency of firm jello. A salt bridge serves three functions.

1.   It allows electrical contact between the two solutions.

2.   It prevents mixing of the electrode solutions.

3.   It maintains electrical neutrality in each half-cell as ions flow into and out of the salt bridge.

Standard electrodes are electrodes in which oxidized and reduced forms of a species are in contact in their thermodynamic standard states. (For our purposes, 1.0 M solutions of ions, and gases at 1.0 atmosphere pressure.) A cell that consists of two standard electrodes is called a *standard cell*.

21-9   The Zinc-Copper Cell

This cell consists of a strip of copper immersed in a 1.0 M copper(II) sulfate, $CuSO_4$, solution and a strip of zinc immersed in 1.0 M zinc sulfate, $ZnSO_4$, solution. A wire and a salt bridge complete the cell. The initial voltage is 1.10 volts. This cell is called the Daniell cell. See Figure 21-6.

The copper electrode gains mass and the concentration of Cu^{2+} ions decreases in the solution around the copper electrode as the cell operates. The zinc electrode loses mass; the concentration of the Zn^{2+} ions increases in the solution around the zinc electrode.

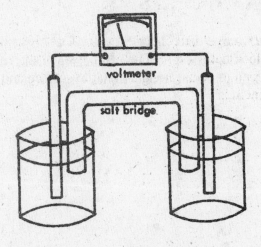

361

Anode reaction

Cathode reaction
_____

Cell reaction

The same net reaction occurs when a piece of metallic zinc is dropped into a solution of $CuSO_4$, but no electricity flows because the oxidation and reduction half-reactions are not separated.

In all *voltaic cells*, electrons flow spontaneously from the negative electrode (anode) to the positive electrode (cathode).

Voltaic cells are frequently represented in shorthand form, as illustrated below for the zinc-copper cell.

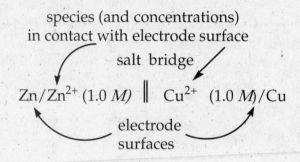

21-10 The Copper-Silver Cell

This cell consists of a strip of copper immersed in 1.0 M copper(II) sulfate solution, and a strip of silver immersed in 1.0 M silver nitrate, $AgNO_3$, solution. A wire and salt bridge complete the circuit. See Figure 21-7.

The copper electrode loses mass; the concentration of Cu^{2+} ions increases around the copper electrode as the cell operates. The silver electrode gains mass; the concentration of Ag^+ ions decreases in the solution around the silver electrode. The *initial* cell potential is 0.462 volt.

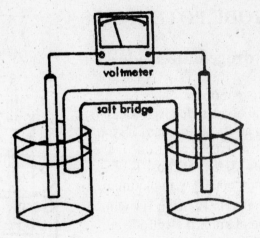

Anode reaction

Cathode reaction
_____

Cell reaction

Recall that in the zinc-copper cell the *copper electrode* was the *cathode*; in the silver-copper cell the *copper electrode* is the *anode*. **Whether a particular electrode behaves as an anode or as a cathode depends on what the other electrode of the cell is.**

The two cells we have described demonstrate that the Cu^{2+} ions are stronger oxidizing agents than Zn^{2+} ions, i.e., Cu^{2+} ions oxidizes metallic zinc to Zn^{2+} ions. By contrast, silver ions are stronger oxidizing agents than copper(II) ions, i.e., Ag^+ ions oxidize copper atoms to Cu^{2+} ions. Zinc metal is a stronger reducing agent than metallic copper, and metallic copper is a stronger reducing agent than metallic silver. We can arrange these species in order of increasing strengths as oxidizing agents and as reducing agents.

strength as oxidizing agents   strength as reducing agents

STANDARD ELECTRODE POTENTIALS

21-11 The Standard Hydrogen Electrode

Because it is not possible to determine experimentally the potential of a single electrode, it is necessary to establish an arbitrary standard. By international agreement the reference electrode is the standard hydrogen electrode (*SHE*). It consists of a piece of metal electrolytically coated with platinum and immersed in a solution that is 1.0 M in H^+ with hydrogen bubbling over the platinized electrode at one atmosphere pressure. The standard hydrogen electrode is arbitrarily assigned a potential of exactly zero volt. When a cell is constructed of a standard hydrogen electrode and another electrode, the cell potential is arbitrarily taken as the potential of the other electrode. See Figure 21-8.

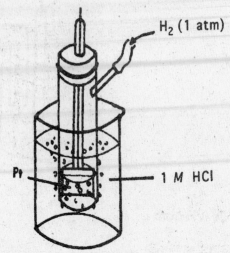

21-12 The Zinc-SHE Cell

This cell consists of a SHE and a strip of zinc immersed in 1.0 M zinc sulfate solution. A wire and a salt bridge complete the circuit. The potential (voltage) is 0.763 volts initially. As the cell operates, the zinc electrode loses mass; the concentration of Zn^{2+} ions increases in the solution around the zinc electrode. The concentration of H^+ decreases in the SHE and hydrogen is produced. See Figure 21-9.

364

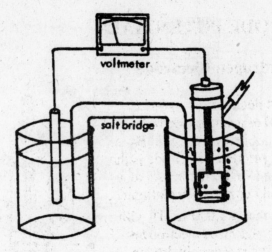

Anode reaction

Cathode reaction _____

Cell reaction

In *this* cell the *SHE is the cathode*, i.e., Zn reduces H^+ ions to H_2. The zinc electrode is the anode, as it was in the zinc-copper cell we examined earlier.

21-13 The Copper-SHE Cell

This cell consists of a standard hydrogen electrode and a strip of copper immersed in 1.0 *M* copper(II) sulfate, $CuSO_4$, solution. A wire and a salt bridge complete the circuit. The initial voltage is 0.337 volts. The copper electrode gains mass; the concentration of Cu^{2+} ions decreases in the solution around the copper electrode. Hydrogen is used up; the concentration of H^+ ions increases in the SHE. See Figure 21-10.

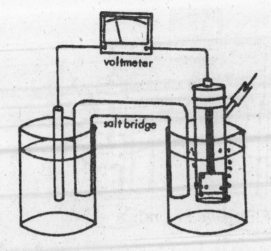

voltmeter

salt bridge

Anode reaction

Cathode reaction

Cell reaction

Note that in this cell the *SHE is the anode*, i.e., Cu^{2+} ions oxidize hydrogen to H^+ ions. The copper electrode is the cathode. In the zinc-SHE cell, the SHE was the cathode.

21-14   Standard Electrode Potentials

Electrodes that force the *SHE to act as the anode* are assigned *positive* standard reduction potentials. In the hydrogen-copper cell, Cu^{2+} ions oxidized H_2 to H^+ ions. The standard reduction potential of the standard copper electrode is +0.337 volt. The oxidized form of the element that forces the SHE to act as an anode is a stronger oxidizing agent than H^+ ions. Or, H_2 is a stronger reducing agent than the reduced species in the other electrode.

Electrodes that force the *SHE to act as the cathode* are assigned *negative* standard reduction potentials. In the zinc-SHE cell, H^+ ions oxidize zinc metal to Zn^{2+} ions. The standard reduction potential of the standard zinc electrode is -0.763 volt.

Let us construct a short table of standard reduction potentials. Table 21-2 in the text provides other examples. A table of standard electrode potentials is referred to as the *electromotive series*, the *EMF series*, or the *activity series*. See Appendix J.

Half-Reaction	E° (volts)

$$2H^+ + 2e^- \rightarrow H_2 \qquad\qquad 0 \text{ (exactly)}$$

21-15 Uses of Standard Electrode Potentials

Standard electrode (reduction) potentials indicate the tendencies of half-reactions to occur as written.

The half-reaction for the standard potassium electrode is

$$E° = -2.925 \text{ V}$$

The very negative E° value tells us that this half-reaction does not occur except under extreme conditions. Electrolysis methods are often utilized to reduce potassium ions to metallic potassium. Potassium is a powerful reducing agent.

The half-reaction for the standard fluorine electrode is

$$E° = +2.87 \text{ V}$$

The very positive E° value tells us that this half-reaction occurs readily. Fluorine, F_2, is the strongest of all common oxidizing agents.

Positive E° values tell us that the tendency of the half-reaction to occur is to the right. The more positive an E° value is, the greater the tendency of the half-reaction to occur to the right. The more negative an E° value is, the greater the tendency of the half-reaction to occur to the left.

Standard electrode potentials and their half-reactions can be used to predict whether a given reaction can occur at thermodynamic standard state conditions [25°C, 1.0 atmosphere partial pressures, and 1.0 M concentrations (actually unit activities)].

Example 21-3:   Will silver ions, Ag^+, oxidize metallic zinc to Zn^{2+} ions, or will Zn^{2+} ions oxidize metallic Ag to Ag^+ ions?

The procedure by which we obtain the equation for the spontaneous reaction follows.

1.   Choose the appropriate half-reactions from a table of standard reduction potentials.

2.   Write the equation for the half-reaction with the more positive $E°$ value first, along with its $E°$ value.

3.   Write the equation for the other half-reaction as *an oxidation* with its *oxidation potential*, i.e., reverse the tabulated reduction half-reaction and change the sign of the tabulated $E°$ value.

4.   Balance the electron transfer. We do not multiply the potentials by the numbers used to balance the electron transfer!

5.   Add the reduction and oxidation half-reactions and their potentials. This produces the equation for the reaction for which $E°_{cell}$ **is positive**, which indicates that the **forward reaction is spontaneous.**

Reduction

Oxidation

Cell reaction

$E°_{cell}$ is positive for this reaction. This tells us that the reaction is spontaneous as written, i.e., Ag^+ ions oxidize Zn to Zn^{2+} ions and are reduced to metallic Ag.

When the procedure we have just illustrated is followed, we always obtain the balanced equation for the spontaneous reaction.

21-16   Standard Electrode Potentials for Other Half-Reactions
(See Table 21-3.)

We can use the same procedures for more complex half-reactions.

Example 21-4:   Will permanganate ions, MnO_4^-, oxidize iron(II) ions to iron(III) ions, or will iron(III) ions oxidize manganese(II) ions to permanganate ions in acidic solution?

Reference to Table 21-3 provides the appropriate half-reactions. We see that the MnO_4^-/Mn^{2+} half-reaction has the more positive $E°$ value, and so we write it first. Then we write the Fe^{3+}/Fe^{2+} half-reaction as an *oxidation*, change the sign of its $E°$ value, balance the electron transfer, and add the half-reactions term by term. **A precautionary note: $E°$ values are *not* multiplied by anything in this procedure.**

Reduction

Oxidation
_____

Cell reaction

We see that permanganate ions, MnO_4^-, oxidize iron(II) ions to iron(III) ions and are reduced to manganese(II) ions in acidic solution.

Example 21-5:   Will nitric acid, HNO_3, oxidize arsenous acid, H_3AsO_3, to arsenic acid, H_3AsO_4, in acidic solution? The reduction product of HNO_3 is NO in this reaction.

Reduction

Oxidation
_____

Cell reaction

The answer is yes, nitric acid oxidizes arsenous acid to arsenic acid.

   Appendix J contains many half-reactions and their $E°$ values.

21-17 Corrosion

   Corrosion of metals refers to oxidation-reduction reactions of a metal with one or more components of the atmosphere such as CO_2, O_2, and H_2O. The most common kind of corrosion is the process by which metals are oxidized by oxygen in the presence of moisture. See Figure 21-11.

21-18   Corrosion Protection

There are several types of corrosion protection.

1.   Plating a metal with a thin layer of a less active (less easily oxidized) metal.

2.   Connecting the metal to a "sacrificial anode", a piece of a more active metal. See Figure 21-13.

3.   Allowing a protective film to form naturally.

4.   Galvanizing, the process in which steel is coated with zinc, a more active metal. See Figure 21-13.

5.   Painting or coating with a polymeric material such as plastic or ceramic.

EFFECT OF CONCENTRATIONS (OR PARTIAL PRESSURES) ON ELECTRODE POTENTIALS

21-19   The Nernst Equation

Standard electrode potentials, designated E°, refer to thermodynamic standard state conditions: one molar solutions of ions, partial pressures of one atmosphere for gases, and solids and liquids in their standard states at 25°C.

Electrode potentials change when concentrations or partial pressures change. If the concentration of ions in a solution is other than 1.0 *M*, or the pressure of a gas is other than one atmosphere, the electrode potential is *not* the standard electrode potential. The Nernst equation allows us to calculate electrode potentials for concentrations and partial pressures other than standard state conditions. The Nernst equation is

E = potential under conditions of interest

E° = potential under thermodynamic standard state conditions

R = universal gas constant = 8.314 J/mol•K

T= absolute temperature in K

n = number of e⁻ transferred

370

F = the Faraday = 96, 487 C/mol e⁻ × (1J/C•V) = 96,487 J/V•mol e⁻

Q = the reaction quotient

Substituting these values into the Nernst equation at 25°C, gives

For a typical half-reaction,

$$Cu^{2+} \; + \; e^- \; \rightleftharpoons \; Cu^+ \qquad\qquad E° = +0.153V$$

The corresponding Nernst equation is

Substituting E° into the above expression gives

Note that when $[Cu^{2+}]$ and $[Cu^+]$ are both 1.0 M, the concentration of solutions at standard state conditions, E = E° because the correction term is zero and E becomes equal to E°.

Because log 1 = 0, we have

Example 21-6:   Calculate the potential for the Cu^{2+}/Cu^+ electrode at 25°C when the concentration of Cu^+ ions is three times that of Cu^{2+} ions.

$$Cu^{2+} \; + \; e^- \; \rightleftharpoons \; Cu^+ \qquad\qquad Q = \frac{[Cu^+]}{[Cu^{2+}]} = \frac{3[Cu^{2+}]}{[Cu^{2+}]} = 3$$

Example 21-7:   Calculate the potential for the Cu^{2+}/Cu^+ electrode at 25°C when the Cu^+ ion concentration is 1/3 of the Cu^{2+} ion concentration.

$$Cu^{2+} + e^- \rightleftharpoons Cu^+ \qquad Q = \frac{[Cu^+]}{[Cu^{2+}]} = \frac{\frac{1}{3}[Cu^{2+}]}{[Cu^{2+}]} = \frac{1}{3}$$

Example 21-8:   Calculate the electrode potential for a hydrogen electrode in which the $[H^+]$ is 1.0×10^{-3} M and the H_2 pressure is 0.50 atmosphere.

$$2H^+ + 2e^- \rightleftharpoons H_2 \qquad Q = \frac{P_{H_2}}{[H^+]^2} = \frac{(0.500)}{(1.0 \times 10^3)^2} = 5.0 \times 10^5$$

21-20   Using Electrochemical Cells to Determine Concentrations.

We can use the Nernst equation to calculate the potential for a cell that consists of two *nonstandard* electrodes.

Example 21-9:   Calculate the initial potential of a cell that consists of an Fe^{3+}/Fe^{2+} electrode in which $[Fe^{3+}] = 1.0 \times 10^{-2}$ M and $[Fe^{2+}] = 0.10$ M, connected to a Sn^{4+}/Sn^{2+} electrode in which $[Sn^{4+}] = 1.0$ M and $[Sn^{2+}] = 0.10$ M. A wire and a salt bridge complete the circuit.

We calculate E^o_{cell} by the usual procedure,

372

Reduction

Oxidation _____

Cell reaction

The Nernst equation that describes this cell is

Substitution of the concentrations of the ions into Q enables us to calculate E_{cell}.

21-21 The Relationship of E^{o}_{cell} to $\Delta G°$ and K

In Section 17-10 we studied the relationship between $\Delta G°$ and the equilibrium constant for a reaction.

There is a simple relationship between $\Delta G°$, the standard Gibbs free energy change, and E^{o}_{cell}, the standard cell potential.

$$F = 96{,}487 \text{ J/V mol e}^{-}$$
$$n = \text{number of e}^{-}$$

Combining the two relationships for $\Delta G°$ relates E^{o}_{cell} to K for a reaction.

Example 21-10: Calculate the standard Gibbs free energy change, $\Delta G°$, at 25°C for the following reaction.

$$Pb(s) \; + \; Cu^{2+} \; \rightleftharpoons \; Pb^{2+} \; + \; Cu(s)$$

373

We use the appropriate half-reactions to calculate E_{cell}^{o}

Now that we know E_{cell}^{o}, we can calculate ΔG°

The negative value for ΔG° (positive value for E_{cell}^{o}) indicates that the reaction is spontaneous as written.

Example 21-11:   Calculate the thermodynamic equilibrium constant for the reaction in Example 21-10 at 25°C.

At equilibrium the concentration of Pb^{2+} ions is 5.0×10^{15} times that of the Cu^{2+} ions!

Example 21-12:   Calculate the Gibbs free energy change, ΔG, and the equilibrium constant at 25°C for the following reaction with the indicated concentrations.

$$Zn(s) \quad + \quad 2Ag^{+}(aq)\,(0.30\,M) \quad \rightarrow \quad 2Ag(s) \quad + \quad Zn^{2+}(aq)\,(0.50\,M)$$

First we calculate the standard cell potential, E_{cell}^{o}

We use the Nernst equation to calculate E_{cell}^{o} for the given concentrations.

$$Q = \frac{[Zn^{2+}]}{[Ag^{+}]^2} = \frac{(0.50)}{(0.30)^2} = 5.6$$

374

$E_{cell} = +1.540$ V, compared to $E_{cell}^o = +1.562$ V. Now we can calculate ΔG.

The negative value for ΔG (this is *not* ΔG°) indicates that the reaction is spontaneous as written (for the given concentrations). The equilibrium constant for a reaction does not change with concentrations of the reactants. To calculate the equilibrium constant, we use the value for E_{cell}^o.

PRIMARY VOLTAIC CELLS

As a voltaic cell produces current (discharges), its chemicals are consumed. Once they are consumed, further chemical action is not possible. Primary voltaic cells cannot be "recharged." This is because the electrodes and electrolytes cannot be regenerated by reversing the current flow through the cell.

21-22 Dry Cells

The most familiar example of a primary voltaic cell is the ordinary dry cell used for flashlight batteries, etc. See Figure 21-15. The container of dry a cell is made of zinc, which serves as one of the electrodes. The zinc container is lined with porous paper that separates the metal from the materials within the cell. The other electrode is a carbon (graphite) rod in the center of the cell. The space between the graphite rod and the zinc container is filled with a moist mixture of ammonium chloride, NH_4Cl, manganese(IV) oxides MnO_2, zinc chloride, $ZnCl_2$, and a porous inactive solid. This cell is also known as the Leclanchè cell. Dry cells are sealed to keep the moisture within the cell.

As the cell operates, zinc dissolves and goes into solution as Zn^{2+} ions, leaving electrons on the zinc container. The zinc electrode is the negative electrode (anode).

The graphite rod is the cathode at which ammonium ions are reduced. Addition of the two half-reactions gives the cell reaction.

Anode

Cathode _____

Cell reaction

As H_2 is formed, it is oxidized by MnO_2. This prevents the collection of H_2 (gas) on the cathode, which would stop the reaction, a condition called *polarization*.

At the cathode, NH_3 combines with Zn^{2+} ions to form a soluble complex compound.

This reaction prevents, polarization due to the accumulation of NH_3 (gas), and it also prevents an increase in the concentration of zinc ions.

The *alkaline dry cell* is similar to the ordinary dry cell except they do not involve the gas production of NH_3. The half-reactions that occur during discharge are

Anode

Cathode _____

Cell reaction $E° = 1.5$ V

Alkaline batteries have longer shelf life and they last longer under heavy use.

SECONDARY VOLTAIC CELLS

In secondary, or reversible cells, the original reactants (electrodes and electrolytes that are consumed in the production of electricity) can be regenerated. This is done by passing a direct current through the cell. The process is "recharging".

21-23   The Lead Storage Battery

The lead storage battery is the most common secondary voltaic cell (Figure 21-17). The electrodes are two sets of lead alloy plates (grids). The holes in one set of grids are filled with lead(IV) oxide, PbO_2, and the holes in the other are filled with spongy lead.

Dilute sulfuric acid is the electrolyte. When the battery operates as a voltaic cell, i.e., delivers current, the spongy lead is oxidized to Pb^{2+} ions, and these plates become negatively charged.

The Pb^{2+} ions combine with SO_4^{2-} ions from H_2SO_4. $PbSO_4$ begins to coat the Pb electrode.

When the *cell delivers current*, the *net* reaction at the *anode* (the lead electrode) is

Electrons are produced at the Pb electrode. They flow through the external circuit to the PbO_2 electrode. In the acidic solution, PbO_2 is reduced to Pb^{2+} ions. These combine with SO_4^{2-} ions; the PbO_2 electrode becomes coated with $PbSO_4$. The PbO_2 electrode is the positive electrode (cathode). When the *cell delivers current* the cathode reaction is

When a lead storage battery *delivers current*, the *cell reaction* may be represented as

Anode

Cathode

Cell reaction

A lead storage battery can be **recharged** by passing electrons, i.e., an external electric current, through the cell in the reverse direction. An external potential converts the cell into an electrolytic cell during the recharging process. The reactions are the reverse of those that occur when the cell operates as a voltaic cell. At the lead electrode, lead ions are reduced to lead atoms. At the lead(IV) oxide electrode, lead(II) ions are oxidized to lead(IV) oxide.

Pb electrode

PbO_2 electrode

Overall reaction

The concentration of H_2SO_4 decreases as the cell discharges. Recharging regenerates H_2SO_4. Consequently, measuring the concentration of the H_2SO_4, usually by determining the density of the solution, is a simple method of measuring the battery's charge.

21-24   The Nickel-Cadmium (Nicad) Cell

The nicad cell can be recharged, and so it is used in electronic calculators, electronic wristwatches, and photographic equipment. The discharge half-reactions are

Anode

Cathode

Cell reaction $E° = 1.4$ V

Because no gases are produced, the cell can be sealed. None of the reactants or products escapes. The cell is recharged by reversing the reaction with an external potential (current).

21-25   The Hydrogen-Oxygen Fuel Cell

Fuel cells are voltaic cells in which reactants are continuously supplied to the cell in the presence of an appropriate catalyst. The hydrogen-oxygen fuel cell is used in space-craft (Figure 21-17). H_2 is oxidized at the anode while O_2 is reduced at the cathode.
Anode

Cathode

Cell reaction

This is just the combination of hydrogen and oxygen. The water produced by the reaction is consumed by astronauts on the spacecraft. The efficiency of energy conversion (50-70%) in the hydrogen-oxygen fuel cell is approximately twice that obtained by burning hydrogen in a heat engine coupled to a generator (and *much greater* than that of internal combustion engines).

Synthesis Question

What were the explosive chemicals in the fuel cell that exploded aboard Apollo 13?

Group Question

Some of the deadliest snakes in the world, for example the cobra, have venoms that are neurotoxins. Neurotoxins have an electrochemical basis. How do neurotoxins disrupt normal chemistry and eventually kill those who have been injected with neurotoxins?

Chapter Twenty Two

METALS I: METALLURGY

METALS

About 75% of the known elements are metals. They are used as structural materials in buildings, vehicles, and systems requiring strength and used also for the conduction of heat and electricity. Many metal ions are essential to health.

22-1   Occurrence of the Metals

Metals with negative standard reduction potentials (active metals) are found in nature in the combined state. These metals are found as ores that must be reduced to give the free metals. Metals with positive standard reduction potentials (less active metals) may occur in the uncombined free state as native ores. In either case the metals are found mixed with soil, clay, sand, and rock (collectively called gangue) that must be removed prior to purification.

METALLURGY

Metallurgy is the commercial extraction of metals from their ores and the preparation of metals for use. Steps that are included in this process are

1.   Mining of the ore

2.   Pretreatment of the ore

3.   Reduction of the ore to the free metal

4.   Refining or purifying the metal

5.   Alloying if necessary

22-2 Pretreatment of Ores

When an ore is mined it is usually mixed with an extensive amount of useless rock or *gangue* taken with the ore which must be removed and the ore concentrated. For ores such as sulfides which have relatively high densities, the ore and gangue are pulverized and then the lighter gangue is separated by blowing it away or sifting it away through wire mesh or on inclined vibration tables.

Another separation method is **floatation**. Ores that are either not "wet" by water, or that can be made water repellent by coating with oil, are mixed with water and oil. These ores rise to the surface of a suspension formed by an air stream blowing through the mixture. A frothy ore concentrate forms at the surface.

Some ores may be changed to more easily reduced forms by chemical modification. Carbonates and hydroxides may be converted to oxides by heating to drive off CO_2 and H_2O (thermal decomposition). For example, calcium carbonate is thermally decomposed into calcium oxide and carbon dioxide while magnesium hydroxide is decomposed into magnesium oxide and water.

Some sulfides are converted to oxides by **roasting**, i.e., heating below their melting points in air. For example, roasting zinc sulfide in air gives zinc oxide and sulfur oxide.

It should be noticed that the production of SO_2 in such processes presents pollution problems that now must be controlled. What use can be made of the SO_2 from such processes?

22-3   Reduction to the Free Metals

How an ore is reduced or smelted to the free metal depends upon the strength of bonding between the metal ions and the anions. When bonding is stronger, more energy is required to reduce the metals, increasing the expense. The most active metals usually have the strongest bonding.

The least active metals such as gold, silver and platinum occur in the free state, and so they require no reduction. Slightly more active metals, such as mercury, can be obtained directly from their sulfide ores by roasting, which yields the metal and SO_2.

Roasting more active metal sulfides, such as zinc sulfide (above) or nickel(II) sulfide, yields metal oxides and SO_2, but no free metals.

The resulting metal oxides are then reduced to free metals with coke (C) or CO. Other reducing agents such as H_2, Fe, or Al are used when C must be avoided. Tin(IV) oxide, SnO_2, is reduced with C; tungsten(VI) oxide, WO_3, is reduced with H_2.

The very active metals such as Al and Na are reduced electrochemically, usually from their molten salts. If water is present, it is reduced in preference to the metal because less energy is required.

22-4 Refining of Metals

Further purification of the impure metals obtained from reduction is usually required. Refining methods include distillation, electrolytic and zone refining. Mercury is distilled because it is more volatile than its impurities. Cu, Ag, Au, and Al are purified electrolytically. The impure metal is the anode and a small piece of the pure metal is the cathode with both electrodes immersed in a solution of the metal's ions.

Zone refining involves a melted zone of the metal moving through a bar of metal. Impurities tend to be carried in the melted zone as the purer metal crystallizes out behind the melted zone. Multiple passes can yield very high purities.

METALLURGIES OF SPECIFIC METALS

22-5 Magnesium

Most magnesium comes from salt brines and from the sea. See Figure 22-5. Magnesium ions are precipitated as $Mg(OH)_2$ by addition of $Ca(OH)_2$ (slaked lime) to sea water. The slaked lime is produced by crushing oyster shells ($CaCO_3$), heating them to produce lime (CaO), and then adding a limited amount of water (slaking).

lime production:

slaking lime:

precipitation:

The precipitation occurs because $Mg(OH)_2$ is much less soluble than $Ca(OH)_2$, its K_{sp} is approximately 10^{-5} times smaller than that of $Ca(OH)_2$. The $Mg(OH)_2$ suspension is filtered and the solid $Mg(OH)_2$ is neutralized with HCl to produce a $MgCl_2$ solution:

384

The $MgCl_2$ is dried and then melted and the melt electrolyzed under an inert atmosphere to produce molten Mg and gaseous Cl_2. The products must be separated as they are formed to prevent recombination:

22-6   Aluminum

Aluminum is obtained from the ore bauxite, a hydrated aluminum oxide, $Al_2O_3 \cdot xH_2O$. Because aluminum ions can be reduced to metal only in the absence of water, the aluminum ore must undergo several transformations to obtain the metal.

$Na[Al(OH)_4]$ is then filtered and neutralized

The solid $Al_2O_3 \cdot xH_2O(s)$ is separated by filtration and heated to form anhydrous Al_2O_3 by heating.

The Al_2O_3 is mixed with cryolite, $Na_3[AlF_6]$, to lower its melting point (a commercial application of freezing point depression) and the melt is electrolyzed using carbon electrodes.

Cathode reaction

Anode reaction

Net reaction

22-7   Iron

Iron may be obtained from the ores hematite, Fe_2O_3, or magnetite, Fe_3O_4, by reduction with carbon monoxide in a blast furnace. Coke is mixed with limestone and crushed ore and then brought into the top of the furnace as the "charge". The carbon monoxide is produced from coke and hot air.

Fe_2O_3 is reduced by carbon monoxide to iron (liquid) and carbon dioxide with evolution of heat.

Some Fe_2O_3 is reduced by coke to iron (liquid). The coke is oxidized to CO. This reaction is also exothermic.

Much of the CO_2 reacts with excess coke to produce more CO.

This CO can reduce the next incoming "charge". A molten slag of calcium silicate is formed by the reaction of the limestone flux with the silica gangue.

The slag floats on the surface of the molten iron, protecting it from atmospheric oxidation. Both slag and iron are drawn off periodically.

22-8 Copper

The main **copper** ores are mixed sulfides of copper and iron such as chalcopyrite, $CuFeS_2$ (or CuS•FeS), and the basic carbonates such as azurite, $Cu_3(CO_3)_2(OH)_2$, and malachite, $Cu_2CO_3(OH)_2$.

To obtain copper from $CuFeS_2$, the compound is separated from gangue by flotation and then roasted to remove volatile impurities. Sufficient air is used to convert the iron(II) sulfide, but not the copper(II) sulfide, to oxide.

The roasted ore is then mixed with sand (silica, SiO_2), crushed limestone ($CaCO_3$), and some unroasted ore that contains copper(II) sulfide in a reverberatory furnace at 1100°C. CuS is reduced to Cu_2S, which melts. The limestone and silica form a molten calcium silicate glass.

This dissolves iron(II) oxide forming a slag.

This slag is less dense than the molten copper(I) sulfide upon which it floats. The slag is periodically drained off. The molten copper(I) sulfide is drawn off into a Bessemer converter where it is heated and air-oxidized to free copper and SO_2.

This oxidizes sulfides, while copper(I) ions are reduced to metallic copper. The impure copper is refined electrolytically as described in Chapter 21.

22-9   Gold

Gold usually occurs in the native state because it is an inactive metal. Panning for gold makes use of gold's high density. Gold bearing sand and gravel are gently swirled with water in a pan. The lighter particles spill over the edge and the denser particles and nuggets of gold remain in the pan.

Gold is also recovered from the anode sludge from the electrolytic refining of copper. Because gold is so rare and valuable, it is also obtained from low-grade ores by the cyanide process. Air is bubbled through an agitated slurry of ore mixed with a solution of NaCN. The gold is slowly oxidized and then forms a soluble complex compound.

After filtration, free gold can then be obtained by reduction of $[Au(CN)_2]^-$ with zinc or by electrolytic reduction.

Synthesis Question

As humans have mastered various metals over the ages, that era of time has been given the metal's name. For example, the brass age and the bronze age were important times of human growth. What metal have we mastered better than any other group of humans? In other words, what metal age do we live in?

Group Question

How much does it cost to refine enough bauxite to make one pound of aluminum? Compare that to the cost of recycling enough aluminum to make one pound of aluminum.

Chapter Twenty Three

METALS II: PROPERTIES AND REACTIONS

The representative metals are those in the A groups of the periodic table. They have valence electrons in their outermost s and p atomic orbitals. Metallic character (increases, decreases) _____ from top to bottom within groups and (increases, decreases) _____ from left to right within periods.

All the elements in Group IA (except H) and 2A are metals. The heavier elements of Groups 3A, 4A, and 5A are called **post-transition metals**.

IA	2A	3A	4A	5A
Li	Be			
Na	Mg	Al		
K	Ca	Ga		
Rb	Sr	In	Sn	
Cs	Ba	Tl	Pb	Bi

THE ALKALI METALS (GROUP IA)
(Table 23-1)

23-1   Group IA Metals: Properties and Occurrence

Because the alkali metals are so reactive, i.e., so easily oxidized, they are not found free in nature. Except for lithium, these metals are soft. The alkali metals have only one valence electron (ns^1). Because the valence electron is in the outermost shell by itself, it is easily removed, i.e., the Group IA metals have low first ionization energies. The ionization energies decrease with increasing size of the alkali metals (going down the column in the periodic table). All the alkali metals form stable 1+ ions. The hydration energies of the alkali metal ions decrease in magnitude with increasing atomic size, because their charge densities (charge/volume) decrease with size. The standard reduction potential includes the combined effects of ionization energy and hydration energy. Thus it may not be surprising that Li has the most negative standard reduction potential, because its high charge density makes it difficult to remove water molecules in the reduction process.

23-2 Reactions of Group IA Metals
(Table 23-2)

The alkali metals are strong reducing agents that form 1+ cations by losing one electron per metal atom. [Reactions of the alkali metals with H_2 and O_2 were discussed in Sections 6-7 and 6-8, reactions with halogens in Section 7-2 and reactions with water in Section 4-9.]

Consider the reaction of lithium with oxygen, a Group VIA nonmetal, to form lithium oxide, Li_2O, an ionic compound.

Sodium reacts with oxygen to form sodium peroxide, Na_2O_2.

The charge on the peroxide ion is _____; and the oxidation number of oxygen is _____.

Potassium and the heavier alkali metals react with oxygen to form superoxides. Potassium reacts with O_2 to form potassium superoxide, KO_2.

The charge on the superoxide ion is _____; the oxidation number of oxygen is _____.

The alkali metals readily react with water to form hydroxides and H_2. Indeed, as we proceed from lithium to sodium, potassium, rubidium, and cesium, these reactions increase in vigor to become quite violent. The reaction of sodium with water is

molecular equation

ionic equation

Diagonal Similarities exist between elements in successive groups near the top of the periodic table.

IA	2A	3A	4A
Li	Be	B	C
Na	Mg	Al	Si

For example, the charge density and electronegativity of Li are close to those of Mg, so Li compounds have some resemblance to those of Mg.

The Group IA metal oxides are basic. They react with water to form strong soluble bases. The reaction of sodium oxide with water is

molecular equation

ionic equation

23-3   Uses of Group IA Metals and Their Compounds

Lithium, Li

Lithium is extremely lightweight and has the highest heat capacity of any element. It is used as a heat transfer medium in experimental nuclear reactors.

Sodium, Na

Sodium is the most widely used alkali metal because of its abundance. Sodium metal is used as a reducing agent. Uses for several Na compounds include

Compound	Use
$NaOH$	
Na_2CO_3 or $Na_2CO_3 \cdot 10H_2O$	
$NaHCO_3$	
$NaNO_3$	
Na_2SO_4	

Other Group IA Metals

Like the salts of sodium, potassium salts are also essential for life. KNO_3 is a fertilizer. There are very few practical uses for rubidium, cesium and francium.

THE ALKALINE EARTH METALS (GROUP 2A)
(Table 23-3)

23-4   Group 2A Metals: Properties and Occurrence

The alkaline earth metals are all silvery white, malleable, ductile, and somewhat harder than Group IA metals. They all have two valence electrons (ns^2). These two electrons are lost in compound formation, although not as easily as the one electron of an alkali metal. Therefore, the ionization energies are greater for Group 2A than for Group IA. While most Group 2A metals are ionic, those of Be show a great deal of covalent character making them somewhat similar to compounds of aluminum in Group 3A. (Recall the diagonal similarities from the previous page.) The Group 2A metals show the +2 oxidation state in all of their compounds. The ease of formation of 2+ ions increases from Be to Ra.

Although the Group 2A metals are less reactive than those of Group IA, they are too reactive to occur free in nature. The metals are prepared by the electrolysis of their molten chlorides.

Calcium and magnesium are quite abundant in the earth's crust, mainly as carbonates and sulfates. Beryllium, strontium, and barium are less abundant. All known radium isotopes are radioactive and extremely rare.

23-5   Reactions of the Group 2A Metals
(Table 23-4)

Reactions of alkaline earth metals are quite similar, with the exception of stoichiometry, to those of the alkali metals. Reactions with hydrogen and oxygen were discussed in Sections 6-7 and 6-8, parts 2.

Except for Be, all the Group 2A metals are oxidized in air to oxides. The 2A oxides (except for BeO) are basic and react with water to give hydroxides. Calcium oxide, CaO, reacts with water to give calcium hydroxide, $Ca(OH)_2$.

molecular equation

ionic equation

$Be(OH)_2$ is quite insoluble and is amphoteric. $Mg(OH)_2$ is only slightly soluble in water. The hydroxides of Ca, Sr, and Ba are strong soluble bases.

$$\longleftarrow$$
Direction of increase in metallic character
and of increase in basicity of oxides
in periodic table
$$\longleftarrow$$

Calcium, Sr, and Ba react with water to form hydroxides and H_2. The equation for the reaction of Ca and water is

Magnesium reacts with steam to give MgO and H_2.

Beryllium does not react with water, but it does dissolve in strong soluble bases by forming the complex ion, $[Be(OH)_4]^{2-}$, and H_2. Balance the reaction expression.

_____ Be + _____ OH^- + _____ H_2O $\rightarrow$ _____ $[Be(OH)_4]^{2-}$ + _____ H_2

23-6   Uses of Group 2A Metals and Their Compounds

Beryllium, Be

Beryllium is quite rare and only has a couple practical uses. One is beryl, a gemstone which, with appropriate impurities may be aquamarine or emerald.

Magnesium, Mg

Metallic magnesium is used in photographic flash accessories and fireworks. This is used because metallic magnesium burns in air with a brilliant white light.

Calcium, Ca

Calcium and its compounds are widely used commercially.

Give formulas for the following compounds of Ca.

quicklime   _____

slaked lime   _____

gypsum   _____

plaster of Paris   _____

Strontium, Sr

The metal strontium itself has no practical uses, but some strontium salts are used in fireworks and flares which give a characteristic red glow.

Barium, Ba

All soluble barium salts are toxic, but BaSO4 is insoluble and is used to coat the gastrointestinal tract in preparation for X-ray photographs because it absorbs X-rays well.

THE POST-TRANSTITION METALS

The metals of Groups 3A, 4A, and 5A are called **post-transition metals**.

3A	4A	5A
Al		
Ga		
In	Sn	
Tl	Pb	Bi

The metals along the stepwise division are really **metalloids**. Aluminum has many properties characteristic of metals. Aluminum it is the only post-transition metal that is considered very reactive.

23-7   Periodic Groups 3A, 4A and 5A: Trends

The properties of the elements in Groups 3A (Table 23-5) show greater irregularities going down the groups than do elements in Groups IA and 2A. All Group 3A elements are solids. Boron, a nonmetal, crystallizes as a covalent solid and thus has a high melting point (2330°C). The other elements of Group 3A form metallic crystals with considerably lower melting points.

Aluminum, Al

Aluminum shows the +3 oxidation state in all its compounds. The heavier metals (Ga, In, Tl) of Group 3A can lose or share either the one p electron or all three valence electrons to show the +1 and +3 oxidation states, respectively. The 4A metals can lose or share the two p electrons to show the +2 oxidation state or all four electrons to give the +4 oxidation state. The stability of the lower oxidation state increases going down the column. This is called the **inert s-pair effect** because the two s electrons remain nonionized or unshared in the lower oxidation state.

Aluminum reacts with nitric acid (if the passive coating of Al_2O_3 is sanded off) to form aluminum nitrate, NO, and water.

molecular equation

ionic equation

In the **thermite reaction**, Al is used as a reducing agent for iron(III) oxide to produce molten iron and aluminum oxide. This spectacular reaction evolves 852 kJ of heat per mole of aluminum oxide produced.

Metallic aluminum is amphoteric as is its hydroxide. Al dissolves in a nonoxidizing acid such as HCl to form aluminum chloride and H_2.

molecular equation

total ionic equation

net ionic equation

Other Group 3A Metals

Gallium has a very unusually low melting point; it melts when held in the hand. Indium is a soft bluish metal, and thallium is a soft, heavy metal.

Periodic Trends

The outer electronic configurations of elements in these groups are

3A	4A	5A
ns^2np^1	ns^2np^2	ns^2np^3

Oxidation states of the post-transition metals are

Al _____

Ga _____

In _____   Sn _____

Tl _____   Pb _____   Bi _____

THE d-TRANSITION METALS

The term "transition elements" applies to elements in the middle of the periodic table that are transitional from the Group IA and 2A base forming elements to the nonmetal acid forming elements. These include the d-transition metals where the d orbitals are being filled and the f-transition elements where the f orbitals are being filled. The f-transition elements, sometimes called inner transition elements, include the rare earths (La-Lu) and the actinides (Th-Lr). The term "transition metals" is commonly used to refer to the d-transition metals.

The d-transition metals occur between Groups 2A and 3A in the periodic table. More precisely, d-transition metals have incompletely filled d orbitals. This means that the elements of Group 2B, zinc, cadmium, and mercury that have filled d orbitals are not exactly d-transition metals. They are usually discussed with d-transition metals because their properties are similar to those of the d-transition metals. All of the elements of this region of the periodic chart have partially filled d orbitals, except the IB elements and palladium. Some of the cations of these latter elements have partially filled d orbitals.

398

23-8   General Properties

Table 23-6 lists some of the properties of the 3d-transition metals. The following are properties of most transition metals.

1.   All are metals.

2.   Most are harder, more brittle, and have higher melting points and higher heats of vaporization than nontransition metals.

3.   Their ions and their compounds are usually colored.

4.   They form many complex ions (Chapter 25).

5.   With few exceptions, they exhibit multiple oxidation states.

6.   Many of the metals and their compounds are paramagnetic.

7.   Many of the metals and their compounds are effective catalysis.

23-9   Oxidation States

The valence electrons of the d-transition metals are their ns electrons and some of their $(n-1)d$ electrons. The properties of these metals vary less dramatically than those of representative elements, whose valence electrons are all in their outer shells. Properties of these metals can be correlated approximately with either the total number of d electrons or the number of unpaired electrons.

For the d-transition metals, the outer s electrons are always the first electrons to be lost in ionization. In the first, or 3d transition series, only Sc and Zn exhibit a single nonzero oxidation state. Zn loses both its 4s electrons to form Zn^{2+}. Sc loses its two 4s electrons and its only 3d electron to form Sc^{3+}.

	3d	4s			3d	4s
$_{30}$Zn [Ar] __ __ __ __ __ __			$\xrightarrow{-2e^-}$	$_{30}$Zn^{2+} [Ar] __ __ __ __ __ __		

	3d	4s			3d	4s
$_{21}$Sc [Ar] __ __ __ __ __ __			$\xrightarrow{-3e^-}$	$_{21}$Sc^{3+} [Ar] __ __ __ __ __ __		

399

All of the other 3d-transition metals show at least two oxidation states in their compounds. For example, iron can form Fe^{2+} and Fe^{3+} ions.

$$_{26}Fe \ [Ar] \ \underline{\quad} \ \underline{\quad} \ \underline{\quad} \ \underline{\quad} \ \underline{\quad} \ \underline{\quad} \quad \xrightarrow{-2e^-} \quad _{26}Fe^{2+} \ [Ar] \ \underline{\quad} \ \underline{\quad} \ \underline{\quad} \ \underline{\quad} \ \underline{\quad} \ \underline{\quad}$$

$$_{26}Fe \ [Ar] \ \underline{\quad} \ \underline{\quad} \ \underline{\quad} \ \underline{\quad} \ \underline{\quad} \ \underline{\quad} \quad \xrightarrow{-3e^-} \quad _{26}Fe^{3+} \ [Ar] \ \underline{\quad} \ \underline{\quad} \ \underline{\quad} \ \underline{\quad} \ \underline{\quad} \ \underline{\quad}$$

(with column labels 3d, 4s over each set of blanks)

The most common oxidation states of the 3d-transition metals are +2, +3 and +4, with +5 and +6 coming next. The elements in the middle of the series show more oxidation states than those on either side of the middle. Going down a Group, higher oxidation states become more stable and more common.

The common oxidation states for the 3d-transition metals (Table 23-7) and their electron configurations are:

	1st Oxidation State	2nd Oxidation State	3rd Oxidation State
Sc			
Ti			
V			
Cr			
Mn			
Fe			
Co			
Ni			
Cu			
Zn			

23-10   Chromium Oxides, Oxyanions, and Hydroxides

Chromium exhibits +2, +3, and +6 oxidation states. Table 23-8 shows some Cr compounds in its several oxidation states. Complete the following table.

Oxidation State	Oxide	Hydroxide	Name	Acidic/ Basic	Related Salt	Name
+2	_____	_____	_____	_____	_____	_____
+3	_____	_____	_____	_____	_____	_____
+6	_____	_____	_____	_____	_____	_____

Oxidation-Reduction

The most stable oxidation state of Cr is +3. Cr^{2+} is a strong reducing agent.

$$Cr^{3+} + e^- \rightarrow Cr^{2+} \qquad\qquad E^o = -0.41 \text{ volt}$$

Chromium(VI) compounds are oxidizing agents. CrO_4^{2-} (which exists in basic solution) is weakly oxidizing. $Cr_2O_7^{2-}$ (which is produced from CrO_4^{2-} by acidification) is a powerful oxidizing agent.

$$Cr_2O_7^{2-} + \_H^+ + \_e^- \rightarrow \_Cr^{3+} + \_H_2O \qquad E^o = +1.33 \text{ volt}$$

Chromate-Dichromate Equilibrium

The interconversion of CrO_4^{2-} to $Cr_2O_7^{2-}$ exists in a pH-dependent equilibrium

$$\_CrO_4^{2-} + \_H^+ \rightleftharpoons \_Cr_2O_7^{2-} + \_H_2O$$

$$K_c = \underline{\hspace{3cm}} = 4.2 \times 10^{14}$$

Dichromate ion and dimanganese heptoxide are isoelectronic and have the same geometry. Sketch the geometry (structure) of $Cr_2O_7^{2-}$ and Mn_2O_7.

Cr(OH)$_3$ is amphoteric (Section 10-6).

$$Cr(OH)_3(s) \ + \ \_\_\_H^+ \ \rightarrow Cr^{3+} \ + \ \_\_\_H_2O \qquad \text{(reaction with acids)}$$

$$Cr(OH)_3(s) \ + \ \_\_\_OH^- \ \rightarrow Cr(OH)_4^- \qquad \text{(reaction with bases)}$$

Molybdenum and tungsten are below chromium in Group 7B. As one goes down the column in the periodic table the MO_3 oxides become less acidic and more stable. Because they are more stable they are weaker oxidizing agents.

Synthesis Question

Many highly colored compounds can be made from the transition metals. Dyes in blue jeans, colored glasses for stoplights, even the red color of blood are derived from transition metal complexes. Why are these compounds so highly colored?

Group Question

The f-transition metals, the lanthanides, do not have quite the commercial viability that the d-transition metals presently have. However, all of us have some lanthanide metals in our homes. They are the phosphorus in color televisions. Why are the lanthanides used for this purpose?

Chapter Twenty Four

SOME NONMETALS AND METALLOIDS

About 20% of the known elements are nonmetals. With the exception of H, they are in the upper-right hand corner of the periodic table. In this chapter the noble gases (Group 8A) and the halogens (Group 7A) will be considered.

THE NOBLE GASES (GROUP 8A)

24-1   Occurrence, Uses, and Properties

The noble gases are very low boiling gases. They can be obtained by the fractional distillation of air. Radon is obtained from the radioactive decay of radium salts. The noble gases constitute less than 1% of the atmosphere (Table 24-1).

Noble Gas	Some Uses
_____	_____
_____	_____
_____	_____
_____	_____
_____	_____
_____	_____

The noble gases are colorless, tasteless, and odorless and all have extremely low melting and boiling points. They are monatomic with only very weak London or Van der Waals forces and therefore have very low melting and boiling points. Because polarizability and interatomic interactions increase with increasing atomic size, their melting and boiling points increase with atomic size (and atomic number).

24-2 Xenon Compounds

Until the 1960's, the Group 8A elements were considered chemically inert. In 1962 Neil Bartlett prepared the first noble gas compounds.

Most noble gas compounds are compounds of Xe. The best characterized are the xenon fluorides. These are xenon difluoride, XeF_2, xenon tetrafluoride, XeF_4, and xenon hexafluoride, XeF_6.

COMPOUND	e⁻ PAIRS AROUND Xe	HYBRIDIZATION AT Xe	GEOMETRY
_____	_____	_____	_____
_____	_____	_____	_____
_____	_____	_____	_____

THE HALOGENS (GROUP 7A)

The elements of Group 7A are known as the **halogens**. Their binary compounds are called **halides**.

24-3   Properties

The elements exist as diatomic molecules. The halogens have high electronegativities with F having the highest electronegativity of any element. Most binary compounds of metals and halogens are ionic. F_2, Cl_2, and Br_2 are strong oxidizing agents while I_2 is a mild oxidizing agent. Conversely, F^-, Cl^-, and Br^- are weak reducing agents while I^- is a mild reducing agent. The fluoride ion is so small (radius = 1.36 Å) it is not easily polarized or distorted by cations. The iodide ion is much larger (radius = 2.16 Å), the iodide ion is much more easily polarized or distorted. Compounds containing I^- ions show greater covalent character than those containing F^- ions. The properties of Cl^- and Br^- ions are intermediate between those of F^- and I^- ions.

Melting and boiling points increase from F_2 to I_2, paralleling their increase in size, polarizability, and interatomic interactions. All halogens except At are nonmetallic. All show the -1 oxidation state in most of their compounds. Except for fluorine, they also show +1, +3, +5, and +7 oxidation states in some compounds.

407

24-4   Occurrence, Production and Uses

The halogens are too reactive to occur free in nature. The most abundant sources of halogens are halide salts. Iodine is usually obtained from $NaIO_3$. The other halogens are obtained by oxidation of the halide ions. The oxidation of Br^- to Br_2 is represented by the following equation.

The order of increasing ease of oxidation of halide ions is

_____ > _____ > _____ > _____

Fluorine

Ores of fluorine are *fluorspar* or *fluorite*, CaF_2, *cryolite*, Na_3AlF_6, and *fluoroapatite*, $Ca_5(PO_4)_3F$. Fluorine, a pale yellow gas, is prepared by electrolysis of KHF_2 under anhydrous conditions in a Monel steel cell. F_2, H_2, and potassium fluoride are produced. The equation for the reaction is

Chlorine

Chlorine occurs in abundance in $NaCl$, KCl, $MgCl_2$, and $CaCl_2$ in salt water and in salt beds. Gastric juices contain HCl. Chlorine, a toxic, yellowish-green gas, is prepared by electrolysis of concentrated aqueous $NaCl$ solution.

Bromine

Bromine occurs mainly in the bromides of Na, K, Mg, and Ca in salt water, underground salt brines, and salt beds. It is less abundant than F_2 and Cl_2. Bromine is prepared by the treatment of hot brine.

Iodine

Iodine is a violet-black crystalline solid with a metallic luster. It has an appreciable vapor pressure at $25°C$. The vapor is violet or purple. Iodine can be obtained from dried seaweed or from $NaIO_3$ impurities in Chilean nitrate ($NaNO_3$) deposits. Iodine is prepared by the reduction of iodate ions with hydrogen sulfite ions, HSO_3^-. This reaction produces iodine, I_2, hydrogen sulfate ions, HSO_4^-, sulfate ions, and water.

Iodine is then purified by sublimation. Elemental chlorine or bromine can be used to displace iodine from iodide salts.

24-5   Reactions of the Free Halogens

The free halogens react with most other elements and many compounds. The most vigorously reactive is F_2, which usually oxidizes other species to their highest oxidation states. Iodine is only a mild oxidizing agent (I^- is a mild reducing agent) and usually does not oxidize substances to high oxidation states.

The general reaction of free halogens with metals is

$$nX_2 \quad + \quad 2M \quad \rightarrow \quad 2MX_n$$

Write equations for:

(1) the reaction of fluorine with potassium

(2) the reaction of chlorine with strontium

(3) the reaction of *excess* bromine with iron

All of the halogens react with Group VA elements to form the trihalides. Write the equation of iodine with phosphorus to form phosphorus triiodide

All of the halogens, except I_2, also react with *most* Group VA elements to form pentahalides. The reactions of F_2 with antimony and bismuth are

All of the halogens react with hydrogen sulfide to form free sulfur and the hydrogen halide. The reaction of Cl_2 with H_2S is

409

A free halogen will displace a halide of a less reactive halogen (one that is lower down in the periodic table), e.g.,

$$F_2 \quad + \quad 2Cl^- \quad \rightarrow \quad 2F^- \quad + \quad Cl_2$$

24-6 The Hydrogen Halides and Hydrohalic Acids

The hydrogen halides are colorless, covalent gases with sharp, irritating odors that dissolve in water to give acidic solutions of the hydrohalic acids.

The hydrogen halides may be prepared by direct combination of the elements. The general (overall) reaction in terms of H_2 and X_2 (where X can be F, Cl, Br, or I) is

The reaction that produces HCl occurs rapidly by a photochemical **chain reaction** when the mixture is exposed to light. The several steps are

initiation step

chain propagation steps

termination steps

The production of HBr from the elements is also a photochemical reaction. The reaction of H_2 and I_2 is very slow, even at high temperatures and with light.

The hydrogen halides may be prepared more safely in the laboratory by the reaction of a metal halide with a nonvolatile, nonoxidizing acid (concentrated H_2SO_4 or H_3PO_4). The more volatile HX evolves as a gas. The reaction of potassium bromide and H_3PO_4 is

Concentrated H_2SO_4 oxidizes HBr to Br_2 and HI to I_2. Most nonmetal halides hydrolyze to produce hydrogen halides and an acid or oxide of the nonmetal. Silicon tetrachloride hydrolyzes to give silicon dioxide and hydrogen chloride.

All of the hydrogen halides react with water to produce **hydrohalic acids** that ionize:

$$H_2O + HX \rightarrow H_3O^+ + X^-$$

Dilute HF is a weak acid ($K_a = 7.2 \times 10^{-4}$); but the other hydrohalic acids are strong acids. Their ionizations are essentially complete. The order of increasing acid strength of the hydrohalic acids is:

_____ << _____ < _____ < _____

Although HF is a weak acid, it etches glass by reacting with silicates such as calcium silicate, $CaSiO_3$, to produce volatile and thermodynamically stable silicon tetrafluoride, SiF_4, CaF_2 and water.

24-7   The Oxoacids (Ternary Acids) of the Halogens

The general formulas of the oxoacid (ternary acids) of the halogens are HXO, HXO_2, HXO_3, and HXO_4. The H is bonded through an O in all these acids.

ACID	Ox. No. of X	LEWIS DOT STRUCTURE	ELECTRONIC GEOMETRY at X	MOLECULAR GEOMETRY	SKETCH
HXO	_____	_____	_____	_____	_____
HXO_2	_____	_____	_____	_____	_____
HXO_3	_____	_____	_____	_____	_____
HXO_4	_____	_____	_____	_____	_____

ACID	Name	SALT	Name
HClO	_____	NaClO	_____
HClO$_2$	_____	NaClO$_2$	_____
HClO$_3$	_____	NaClO$_3$	_____
HClO$_4$	_____	NaClO$_4$	_____

Arrange these acids in order of increasing acid strength.

_____ > _____ > _____ > _____

Arrange these acids in order of increasing thermal stability.

_____ > _____ > _____ > _____

Arrange these acids in order of increasing oxidizing power.

_____ > _____ > _____ > _____

Hypohalite salts of Cl, Br, and I may be prepared by reaction of the halogens with **cold** dilute bases. The reaction of bromine with KOH produces potassium bromide and potassium hypobromite.

Aqueous hypohalous acids (except HOF) may be prepared by the reaction of free halogens with cold water. For example, chlorine reacts with cold water to produce HCl and HClO.

Hypochlorous acid decomposes into HCl and oxygen *radicals*.

$$\text{HOCl} \quad \rightarrow \quad \text{HCl} \quad + \quad :\overset{..}{\underset{..}{O}}\cdot$$

Oxygen radicals are strong oxidizing agents. They are the effective bleaching and disinfecting agent in aqueous solutions of Cl$_2$ and hypochlorite salts.

SULFUR, SELENIUM, AND TELLURIUM

24-8   Occurrence, Properties, and Uses

In addition to oxygen, sulfur, selenium, and tellurium (S, Se, and Te) are members of Group 6A (as is polonium, Po). Oxygen was discussed in Section 6-8. The members increase in metallic character going down the group. Oxygen and sulfur are nonmetals, selenium less so. Tellurium is a metalloid and it forms metal-like crystals. The chemistry of tellurium is mostly that of a nonmetal. Polonium is a metal. All its isotopes are radioactive.

The outer electron configuration of Group 6A elements is ns^2np^4. Sulfur and the heavier elements have low energy d orbitals (nd^0) that oxygen does not have. Therefore, the properties of oxygen are not very similar to those of other Group 6A elements. All of the 6A elements may gain or share two electrons in forming compounds. All form covalent compounds of the type H_2E in which the 6A element (E) has a -2 oxidation number. Oxygen has only four orbitals available for bonding and thus can bond to a maximum of four atoms (coordination number of four), but S, Se, Te and probably Po can use one or two of their vacant d orbitals in addition to their s and p orbitals for bonding. They can form up to six bonds.

Sulfur

Sulfur occurs (in nature) as

Formula	Name	Oxidation
_____	_____	_____
_____	_____	_____
_____	_____	_____
_____	_____	_____

Molecular forms of sulfur: The most stable solid forms of sulfur consist of

S_n molecules (where n = _____).

These are puckered rings with all S-S single bonds. At 444°C, sulfur boils to give a vapor that contains the following species.

_____, _____, _____, and _____

413

Selenium

Selenium is quite rare. It occurs mainly as an impurity in sulfur, sulfide and sulfate deposits. It is obtained as a byproduct from roasting sulfide ores and from the "anode mud" in the electrolytic refining of copper.

Tellurium

Tellurium occurs mainly in sulfide ores, particularly with copper sulfide, and as gold and silver telluride.

24-9 Reactions of Group 6A Elements

The group 6A nonmetals react with many metals to form salts. Sulfur reacts with iron to form iron (II) sulfide.

The group 6A nonmetals will react with some of their salts to add additonal group 6A nonmetals: sulfur will react with iron(II) sulfide to form iron(III) sulfuide, Fe_2S_3.

The group 6A nonmetals react with H_2 to give the hydrides, H_2E.

The group 6A nonmetals (S, Se, and Te) react with excess F_2 to form EF_6.

Review table on page 889.

24-10 Hydrides of Group 6A Elements

All of the 6A elements form covalent hydrides of the general formula H_2E, i.e., H_2O, H_2S, H_2Te, and H_2Po. In these compounds, the 6A elements are in the -2 oxidation state. The hydrides of S, Se, Te, and Po are colorless, noxious, poisonous gases, whereas water is a liquid essential to animal and plant life.

The melting point and boiling point of water, H_2O, are primarily determined by hydrogen bonding. No significant H-bonding occurs in the other members of the series in which the electronegativity differences between H and E is much smaller than that of H_2O.

414

Aqueous solutions of the Group 6A hydrides are acidic with the order of increasing acidity:

_____ < _____ < _____

Note that these acids ionize in two steps.

	H_2S	H_2Se	H_2Te

$H_2E \rightleftharpoons H^+ + HE^-$ $\quad K_1$: _____ _____ _____

$HE^- \rightleftharpoons H^+ + E_2^-$ $\quad K_2$: _____ _____ _____

24-11 Group VIA Oxides

The most important oxides are the dioxides, EO_2, acid anhydrides of H_2EO_3, and the trioxides, EO_3, acid anhydrides of H_2EO_4. SO_2 and SO_3 will be used as examples

Sulfur Dioxide, SO_2

Given that the SO bonds in sulfur dioxide are of equal length, draw the electron dot structure(s).

The electronic geometry around S is _____

The molecular geometry around S is _____

When an ore such as zinc sulfide is roasted (combustion in air) sulfur dioxide is produced as a by-product along with the metal oxide. The reaction is:

A way of removing most sulfur dioxide involves the injection of limestone into the combustion zone of the furnace. Limestone will decompose to lime. The reaction is:

The lime will then combine with sulfur dioxide to form calcium sulfite. The reaction is:

Sulfur Trioxide, SO3

Given that the S-O bonds in SO_3 are all of equal length. Draw electron dot structures for SO_3.

The electronic geometry around S is _____.

The molecular geometry around S is _____.

Sulfur trioxide is the anhydride of sulfuric acid. It is made by oxidizing SO_2.

$$2SO_2(g) + O_2(g) \xrightarrow{\text{catalyst}} 2SO_3(g) \qquad \Delta H° = \text{-197.6 kJ}, \qquad \Delta S° = \text{-188J}$$

High temperatures favor SO_2 and O_2. Why is the reaction carried out at high temperatures?

24-12   Oxoacids of Sulfur

Sulfurous Acid, H2SO3

Solutions of sulfurous acid are produced by dissolving sulfur dioxide in water.

Sulfurous acid ionizes in two steps in water.

$$K_1 = 1.2 \times 10^{-2}$$
$$K_2 = 6.2 \times 10^{-8}$$

Example 24-1:   Draw the electron dot formula of the sulfite ion, SO_3^{2-}. What is the electronic geometry, the molecular geometry, and the hybridization at S for SO_3^{2-}.

416

Sulfuric Acid, H$_2$SO$_4$

Sulfuric acid is one of the most important chemicals produced; the annual worldwide production is in excess of 40 million tons.

Sulfuric acid, H$_2$SO$_4$, ionizes in two steps; it is a strong acid with regard to its first ionization step (K$_1$ very large), its second step occurs to a lesser extent.

$$K_1 \approx \infty$$
$$K_2 = 1.2 \times 10^{-2}$$

Sulfuric acid is a colorless, oily liquid that freezes at 10.4°C and boils at 290 to 317°C. A lot of heat is given off when sulfuric acid is diluted and is commonly used as a dehydrating agent.

NITROGEN AND PHOSPHORUS

The Group 5A Elements, also known as the nitrogen family, includes nitrogen and phosphorus which are nonmetals, arsenic which is mainly nonmetallic, antimony which is more metallic, and bismuth which is definitely metallic. Table 24-7 describes the properties of the Group 5A Elements.

The electron configuration of these elements is ns^2np^3. They show oxidation states from -3 to +5. They show the -3 oxidation state in the covalent hydrides, such as NH$_3$, PH$_3$, and AsH$_3$. The +5 oxidation state occurs only in covalent compounds such as phosphorus pentafluoride PF$_5$ and nitric acid, HNO$_3$. The lighter elements, N and P, show many oxidation states in their compounds but the common oxidation states for the heavier elements, As, Sb, and Bi, are +3 and -5. Table 24-8 gives examples of the nitrogen oxidation states.

All Group 5A elements form oxides in which they exhibit the +3 oxidation state: N$_2$O$_3$, P$_4$O$_6$, As$_4$O$_6$, Sb$_4$O$_6$, and Bi$_2$O$_3$. The first two are anhydrides of the weak acids nitrous acid, HNO$_2$, and phosphorus acid, H$_3$PO$_3$; AS$_4$O$_6$, and Sb$_4$O$_6$ are amphoteric with Sb$_4$O$_6$ being more basic. Neither of these last two oxides dissolves in water to any appreciable extent. The trend going down the family is for the element to become more metallic and their oxides to become ore basic.

24-13   Occurrence of Nitrogen

Nitrogen is a colorless, odorless, tasteless gas and makes up about 75% by mass and 78% by volume of the atmosphere. The *nitrogen cycle* is a complex series of reactions by which nitrogen is slowly but continually recycled in the atmosphere, lithosphere, and hydrosphere. NO is also formed by direct reaction of N$_2$ and O$_2$ in electric storms. NO is paramagnetic with one unpaired electron. NO dimerizes in the solid state to N$_2$O$_2$

24-14   Hydrogen Compounds of Nitrogen

Review Section 17-7 and 18-4. *Liquid* ammonia is sometimes used as a polar nonaqueous solvent. Ammonia is hydrogen bonded as is water, but it is a more basic solvent. It undergoes autoionization in a fashion similar to that of water to produce NH_4^+ and NH_2^- ions.

Many ammonium salts are known and most are very soluble in water. Ammonium salts can be prepared by reaction of ammonia with acids. The reaction with nitric acid gives ammonium nitrate.

Amines are organic compounds that are structurally related to ammonia. They thought as derived from NH_3 where one or more of their hydrogens are replaced with an organic group. They are all weak bases.

Species	Lewis Structure	Electronic Geometry at N	Molecular Geometry	Structure
NH_3				
NH_4^+				

24-15   Nitrogen Oxides

There exist several oxides of nitrogen with nitrogen showing oxidation states from +1 to +5. All of these oxides have $\Delta G_f^o > 0$ because of the large dissociation energies of N_2 and O_2 molecules.

Compound	Lewis Structure	Electronic Geometry at N	Molecular Geometry	Structure
N_2O				
NO				
N_2O_2				
N_2O_3				
NO_2				
N_2O_4				
N_2O_5				

Dinitrogen Oxide (+1 Oxidation State)

Molten ammonium nitrate decomposes at 170° to 260°C to produce dinitrogen oxide, also called nitrous oxide.

At higher temperatures ammonium nitrate explosively decomposes into N_2, O_2, and H_2O.

Nitrogen Oxide (+2 Oxidation State)

The commercial preparation of NO is by the catalytic oxidation of ammonia.

NO reacts with O_2 rapidly to form NO_2 (a component of smog in dry climates).

Nitrogen Dioxide and Dinitrogen Tetroxide (+4 Oxidation State)

Nitrogen dioxide, NO_2, a brown corrosive gas, is prepared in the laboratory by heating heavy metal nitrates, e.g., $Pb(NO_3)_2$ to obtain NO_2, O_2 and PbO.

The NO_2 molecule has one unpaired electron and is represented by several resonance structures.

When NO_2 is cooled it easily dimerizes to N_2O_4.

Nitrogen Oxides and Photochemical Smog (Chemistry In Use, The Environment)

Although nitrogen oxides are produced in the atmosphere by natural processes, they are produced in urban areas in much higher concentrations as by-products of human activities. The oxide NO is produced by the reaction of N_2 and O_2 in combustion engines and furnaces and released into the atmosphere. NO reacts with O_2 to make NO_2.

Both of these compounds, NO and NO_2 are quite reactive and cause considerable harm to animals and plants. NO_2 reacts with H_2O in the air to make NO and corrosive droplets of HNO_3.

The pollution that occurs in urban centers is made worse in cities located in warm dry climates favorable to photochemical (light induced) reactions. During the morning rush hour, NO from engine exhaust, is released to the atmosphere. The NO reacts with O_2 to produce NO_2. The NO_2 absorbs ultraviolet light provided by the midday sun to break down to NO and oxygen radicals.

The very reactive O radicals react with O_2 to make ozone, O_3.

Ozone is a powerful oxidizing agent that damages rubber, plastic, plant and *all* animal life.

24-16 Some Oxoacids of Nitrogen and Their Salts

The dominant oxoacids of nitrogen are nitrous acid, HNO_2, and nitric acid, HNO_3.

Acid or Ion	Lewis Structure	Electronic Geometry at N	Molecular Geometry	Structure
HNO_2				
NO_2^-				
HNO_3				
NO_3^-				

Nitrous Acid (+3 Oxidation State)

The anhydride of HNO_2 is _____

Resonance structures of NO_2^-

Nitric Acid (+5 Oxidation State)

The anhydride of HNO_3 is _____

Resonance structures of HNO_3

Resonance structures of NO_3^-

The commercial production of HNO_3 is by the Ostwald process. NH_3 is catalytically oxidized to NO at high temperature

The NO is cooled and then air oxidized to NO_2

The NO_2 reacts with H_2O to make HNO_3 and NO

The NO made in the third step is then recycled into the **second** step. More than 15 billion pounds of nitric acid were made in the United States in 1990.

Nitric acid is quite soluble in water (ca. 16 mol/L). HNO_3 is a strong acid and a strong oxidizing acid.

NaNO$_2$ and NaNO$_3$ as Food Additives

Nitrites and nitrates are added to food to retard the oxidations of blood and to prevent growth of bacteria. Nitrate ions, NO$_3^-$, are reduced to NO$_2^-$ ions which are then converted to NO. This reaction keeps meat red for a longer period of time. It is believed that these nitrates combine with amines under acidic conditions in the stomach to produce carcinogenic *nitrosamines*.

24-17 Phosphorous

Phosphorus only occurs in combined forms in nature. The industrial preparation of phosphorus involves heating phosphate minerals to 1200 to 1500°C in an electric furnace with sand and coke.

$$\_Ca_3(PO_4)_2 + \_SiO_2 + \_C \rightarrow \_CaSiO_3 + \_CO + \_P_4$$

Phosphorus exists in two allotropic forms: white phosphorus, tetrahedral P$_4$ molecules (mp = 44.2°C; bp = 280.3°C) is stored under water to prevent oxidation. Even while under water, white P slowly converts to the more stable red phosphorus which has a polymeric lattic in which one bond in each P$_4$ tetrahedron has been broken and replaced by a bond *between* tetrahedra (mp = 597°C; sublimes at 431°C). Both allotropic forms of P are insoluble in water.

Red phosphorus and tetraphosphorus trisulfide, P$_4$S$_3$, are used in matches. Although they do not burn spontaneously they are ignited easily when heated by friction.

The greatest use of phosphorus is in fertilizers.

SILICON

Silicon is a high-melting, brittle metalloid with a shiny, blue-gray, metallic appearance. Its chemical behavior is more like that of a nonmetal. The earth's crust is 26% Si which occurs in silica, SiO$_2$, and silicates. Si does not occur free in nature. Pure Si crystallizes in a diamond-like structure, but not as closely packed as C in diamond. The densities of the two materials are Si, 2.4 g/cm^3 and diamond (C), 3.51 g/cm^3.

24-18   Silicon and the Silicates

Silica reacts slowly with strong bases to yield various soluble silicates

$$\_SiO_2 + \_NaOH \rightarrow \_Na_2SiO_3(aq) + \_H_2O$$
sodium metasilicate

$$\_SiO_2 + \_NaOH \rightarrow \_Na_4SiO_4(aq) + \_H_2O$$
sodium orthosilicate

$$\_SiO_2 + \_NaOH \rightarrow \_Na_6Si_2O_7(aq) + \_H_2O$$
sodium pyrosilicate

The parent acids of these salts are unstable and dehydration converts them to SiO_2. *Silica gel* is a spongy form of silica containing 5% water. Because it has a large surface area, it absorbs water and some vapors in relatively large quantities. Because of this property it is used as a drying agent.

Glass is a hard, brittle, silicate material that has no fixed composition or regular structure. Because it has no regular structure, it does not fracture or break along crystal planes, but breaks with jagged edges and rounded surfaces. The principal constituents are made by heating a mixture of Na_2CO_3 and $CaCO_3$ with sand until it melts at about 700°C.

$$Na_2CO_3 + SiO_2 \rightarrow$$

$$CaCO_3 + SiO_2 \rightarrow$$

Synthesis Question

Coal which has a high sulfur content cannot be burned in the United States because it is considered to be a serious air pollutant. What pollution effect are we trying to stop with this ban?

Group Question

Is there an economical method to trap the sulfur oxides that are produced in the burning of sulfur rich coal? This would allow us to burn the relatively abundant high sulfur coal that we have in the United States.

Chapter Twenty Five

COORDINATION COMPOUNDS

25-1   Coordination Compounds

A *coordinate covalent bond* is a pair of electrons from a donor shared with an acceptor. Such bonds occur in Lewis acid-base reactions. An example of a coordinate covalent bond is the one formed between ammonia and boron trifluoride.

$$H_3N: \; + \; BF_3 \; \rightarrow \; H_3N:BF_3$$

The Lewis base is _____; the Lewis acid is _____.

The boron of BF_3 has an empty valence orbital, which is able to accept the electron pair donated by NH_3.

Most d-transition metal ions have vacant d orbitals that can accept shares in electron pairs. Many transition metal ions act as Lewis acids by forming coordinate covalent bonds in coordination compounds (coordination complexes or complex ions). Some complexes are $[Cu(NH_3)_4]^{2+}$, $[Co(CN)_6]^{3-}$, $[Fe(CO)_5]$, and $[Ag(NH_3)_2]^+$. Many complexes are quite stable as is indicated by their low dissociation constants, K_d (Appendix I).

The charge on a complex is the sum of its constituent charges (or oxidation states).

The charge on $[Cu(NH_3)_4]^{2+}$ can be calculated as

428

The charge on $[Co(CN)_6]^{3-}$ can be calculated as

25-2   The Ammine Complexes

The *ammine complexes* contain NH_3 molecules bonded to metal ions by coordinate covalent bonds, e.g., $[Cu(NH_3)_4]^{2+}$. Because most metal hydroxides are insoluble in water, dilute aqueous NH_3 reacts with them to form the insoluble metal hydroxides or hydrated oxides. The exceptions are those metals that form strong soluble hydroxides (Group IA cations and the heavier Group 2A cations, Ca^{2+}, Sr^{2+}, Ba^{2+}). For example,

$$Cu^{2+} \quad + \quad NH_3 \quad + \quad H_2O \quad \rightarrow$$

$$Fe^{3+} \quad + \quad NH_3 \quad + \quad H_2O \quad \rightarrow$$

Several metal hydroxides dissolve in excess aqueous NH_3 to form ammine complexes.

$$Cu(OH)_2(s) \quad + \quad 4\,NH_3 \quad \rightarrow$$

$$Co(OH)_2(s) \quad + \quad 6\,NH_3 \quad \rightarrow$$

Some metal ions that form soluble ammine complexes with an excess of aqueous NH_3 include Co^{2+}, Co^{3+}, Ni^{2+}, Cu^{2+}, Ag^+, Zn^{2+}, Cd^{2+}, Hg^{2+}. (See Table 25-3).

25-3   Important Terms

Ligand	A Lewis base that coordinate to a central metal atom or ion.
Donor atom	The atom in a ligand that donates a lone pair of electrons to form a coordinate covalent bond.
monodentate ligand	A ligand that can bind through only one atom.
Polydentate ligand	A ligand that can bind through more than one donor atom. Polydentates that bind through two, three, four, five or six donor atoms are called *bidentate, tridentate, quadridentate, quinquedentate,* and *sexidentate,* respectively.
Chelate complexes	Complexes that have a metal atom or ion and polydentate ligand(s) that form rings.

429

Coordination number	The number of *donor atoms* coordinated to a metal atom or ion.
Coordination sphere	Includes the metal atom or ion and the ligands coordinated to it. Does not include uncoordinated counter ions.

Typical Simple Ligands (See Table 25-4)

Ion/Molecule	Name	Name as Ligand
NH_3	_____	_____
CO	_____	_____
Cl^-	_____	_____
CN^-	_____	_____
F^-	_____	_____
OH^-	_____	_____
NO	_____	_____
NO_2^-	_____	_____
PH_3	_____	_____
OH_2	_____	_____

Example 25-1:   For the complex compound $K_3[Co(CN)_6]$,

the coordination number is _____,

and the coordination sphere is _____.

25-4   Nomenclature

The function of nomenclature is to permit unambiguous communication involving compounds and ions. Rules for naming complex species follow.

1. Cations are named before anions.

2. Coordinated ligands are named in alphabetical order. The prefixes that specify the number of each kind of ligand (di = 2, tri = 3, tetra = 4, penta = 5, hexa = 6, etc.) are not used in alphabetizing. But when a prefix is part of the name of ligand as in diethylamine, it is used to alphabetize the ligands. For complicated ligands, especially those that have a prefix such as di or tri as part of the ligand name, other prefixes are used to specify the number of those ligands (bis = 2, tris = 3, tetrakis = 4, pentakis = 5, and hexakis = 6).

3. The names of most neutral ligands end in the suffix -o. Examples are: Cl^-, chloro; S^{2-}, sulfido; O^{2-}, oxo; OH^-, hydroxo; CN^-, cyano; NO_3^-, nitrato; SO_4^{2-}, sulfato.

4. The names of most neutral ligands are unchanged. However, several important exceptions are NH_3, ammine; H_2O, aqua; CO, carbonyl; and NO, nitrosyl.

5. The oxidation number of a metal that exhibits variable oxidation states is designated by a Roman numeral in parentheses following the name of the complex ion or molecule.

6. If a complex is an anion, the sufix "ate" ends the name. No suffix is used in the case of a neutral or cationic complex. Usually, the English stem is used for a metal, but if this would make the name awkward, the Latin stem is substituted. E.g., ferrate is used instead of ironate, plumbate instead of leadate.

Name the following compounds:

$Na_3[Fe(Cl)_6]$ _____

$[Ni(NH_3)_4(OH_2)_2](NO_3)_2$ _____

Write formulas for the following compounds:

potassium hexacyanochromate(III) _____

tris(ethylenediammine)cobalt(III) nitrate _____

25-5   Structures

The structures of coordination compounds are controlled primarily by the coordination number of the metal.

Usually the structures can be predicted by VSEPR theory (Chapter 8). The geometries and hybridizations for common coordination numbers are summarized below.

CN	Geometry	Hybridization	Example
2	linear	sp	
4	tetrahedral	sp^3	
4	square planar	dsp^2 or sp^2d	
5	trigonal bipyramidal	dsp^3	
5	square pyramidal	d^2sp^2	
6	octahedral	d^2sp^3 or sp^3d^2	

Sketch the shape of the hexacyanaochromate(III) ion

ISOMERISM IN COORDINATION COMPOUNDS

Isomers are compounds that have the same number and kind of atoms but their atoms are arranged differently. Isomers have different properties because they have different structures. There are two major classes of isomers: structural isomers and stereoisomers. With coordination compounds, these classes can be categorized further.

25-6   Structural (Constitutional) Isomers

Structural isomers involve different either more than one coordination sphere or different donor atoms on the same ligand. They contain different atom-to-atom bonding sequences.

Ionization (Ion-Ion Exchange) Isomers

These isomers result from the interchange of ions inside and outside the coordination sphere.

Example: $[Pt(NH_3)_4Cl_2]Br_2$ and $[Pt(NH_3)_4Br_2]Cl_2$

Hydrate Isomers

Hydrate isomers are essentially a special case of ionization isomers in which water molecules may be changed from inside to outside the coordination sphere.

Examples: $[Cr(OH_2)_6]Cl_3$, $[Cr(OH_2)_5Cl]Cl_2 \cdot H_2O$, $[Cr(OH_2)_4Cl_2]Cl \cdot 2H_2O$

Coordination Isomers

Coordination isomers involve exchange of ligands between the coordination spheres of cation and anion.

Examples: $[Pt(NH_3)_4][PtCl_6]$ and $[Pt(NH_3)_4Cl_2][PtCl_4]$

Linkage Isomers

Linkage isomers involve ligands that bind to the metal in more than one way. Examples of this are cyano. $-CN^-$, and isocyano $-NC^-$; nitro, $-NO_2^-$, and nitrito, $-ONO^-$.

Examples: $[Co(NH_3)_5ONO]Cl_2$ and $[Co(NH_3)_5NO_2]Cl_2$

25-7　Stereoisomers

Isomers that have different spatial arrangements of the atoms relative to the central atom are **stereoisomers**. Complexes with only *simple* ligands can occur as stereoisomers. Stereoisomers must have coordination numbers equal to or greater than four.

Geometric (*cis-trans*) Isomers

Geometrical isomers or positional isomers are stereoisomers that are not optical isomers. *Cis-trans* isomers have the same kind of ligand either adjacent to each other (*cis*) or on the opposite side of the central metal atom from each other (*trans*).

cis-diamminedichloroplatinum(II)　　　　　　*trans*-diamminedichloroplatinum(II)

cis- $[Pt(NH_3)_2Cl_2]$　　　　　　　　　　*trans*-$[Pt(NH_3)_2Cl_2]$

Other types of isomerism can occur in octahedral complexes. For example, complexes of the type $[MA_2B_2C_2]$ can occur in several isomeric forms in which *cis* and *trans* isomers are placed in different positions. Examples

trans-diammine-*trans*-diaqua-　　　　　*cis*-diammine-*cis*-diaqua-
trans-dichlorocobalt(III) ion　　　　　　*cis*-dichlorocobalt(III) ion

trans-diammine-*cis*-diaqua-　　　　　　*cis*-diammine-*cis*-diaqua-
cis-dichlorocobalt(III) ion　　　　　　　*cis*-dichlorocobalt(III) ions

434

cis-diammine-*trans*-diaqua-
cis-dichlorocobalt(III) ion

Optical Isomers

The *cis*-diammine-*cis*-diaqua-*cis*-dichlorocobalt(III) ion can exist in two different forms that although mirror images of each other are not superimposable. These are called **optical isomers** or **enantiomers**. The optical isomers of *cis*-diammine-*cis*-diaqua-*cis*-dichlorocobalt(III) ion are

Separate equimolar solutions of the two isomers rotate plane polarized light by equal angles but in opposite directions. This phenomenon of rotation of polarized light is called **optical activity.**

BONDING IN COORDINATION COMPOUNDS

Bonding theories are used to account for structural features and physical properties. The earliest accepted theory was the valence bond theory (Chapter 8). It can account for most structural and magnetic properties of coordination compounds. The **crystal field theory** gives quite satisfactory explanations of color, magnetic properties, and together with valence bond theory, can explain the occurrence of **inner** and **outer** orbital complexes.

25-8   Crystal Field Theory

Crystal field theory in its simplest form treats the ligands as point charges and considers the effect of these point charges on the relative energies of the *d* orbitals. The five *d* orbitals can be divided into two subsets. The d_{z^2} and $d_{x^2-y^2}$ often called the e_g orbitals; they are directed along the x, y, and z axes. The d_{xy}, d_{xz}, and d_{yz} orbitals are called the t_{2g} orbitals; they are directed between the x, y, and z axes.

In octahedral coordination, the ligands approach the central metal along the x, y, and z axes, creating a more repulsive environment for electrons in the e_g orbitals than for electrons in the t_{2g} orbitals. Thus the electric field (crystal field) splits the degeneracy of the five d orbitals into two higher energy orbitals (e_g) and three lower energy orbitals (t_{2g}).

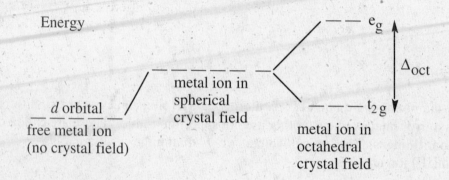

The energy separation between the two sets of d orbitals is the crystal field splitting energy and is also called $\Delta_{octahedral}$ or Δ_{oct}. It is proportional to the *crystal field strength* of the ligands. Common ligands have been arranged in the order of increasing crystal field strengths.

$$I^- < Br^- < Cl^- < F^- < OH^- < H_2O < (COO)_2^{2-} < NH_3 < en < NO_3^- < CN^-$$

--increasing crystal field strength-->

25-9   Color and the Spectrochemical Series

Such an arrangement of ligands is often called a spectrochemical series because it is obtained from the colors or visible spectra of many complexes. The ligand field strength is proportional to the crystal field splitting. Strong field ligands such as CN^- cause large crystal field splitting, and wherever possible low spin complexes. The large splitting provides the energy to cause the electrons to pair in the lower energy d orbitals. Weak field ligands, such as Cl^-, cause only a small crystal field splitting and high spin complexes. Reconsider now our earlier examples of $[Fe(Cl)_6]^{4-}$ and $[Fe(CN)_6]^{4-}$.

$X = Cl^-$ $X = CN^-$

— — e_g

$[FeX_6]^{4-}$ — — — — —

— — e_g

— — — t_{2g}

— — — t_{2g}

d orbital no field   weak field   strong field

We see that $[Fe(Cl)_6]^{4-}$ is a high spin complex and i.e., $3d$ orbitals are occupied and unavailable for hybridization. It is an outer orbital complex using the $4d$ orbitals instead of the $3d$ orbitals to form its bonding hybrids - sp^3d^2. The $[Fe(CN)_6]^{4-}$ complex with a large crystal field splitting is a low spin, inner orbital complex (d^2sp^3).

Metal ion configurations of d^4, d^5, d^6, and d^7 can have different configurations in weak field (high spin) and strong field (low spin) complexes. For d^1-d^3 and d^8-d^{10}, only one configuration (high spin) exists.

Example 25-2:   Name the compound $K_4[MnFe_6]$. _____

What are its geometry, magnetic properties, and hybridization at Mn?

Example 25-3:   Name the compound $[Mn(NH_3)_6]Cl_2$._____

What are its geometry, magnetic properties, and hybridization at Mn?

Synthesis Question

How is it determined experimentally that a transition metal compound contains one, two, three or more unpaired electrons?

438

Group Question

At the center of the heme molecule, a component of red blood cells, is an Fe^{2+} ion. What is the hybridization of the Fe^{2+} ion in heme. Does heme form a high spin or low spin complex?

Chapter Twenty Six

NUCLEAR CHEMISTRY

Nuclear reactions involve changes in the composition or structure of nuclei. This is in distinction to chemical reactions, which involve changes in the extranuclear structure (electrons) of atoms and molecules. A comparison of nuclear reactions and ordinary chemical reactions follows:

Nuclear reaction	**Ordinary chemical reaction**
1. Elements may be converted from one to another.	1. No new elements can be produced.
2. Particles within the nucleus are involved.	2. Usually only the outermost electrons participate.
3. Release or absorb immense amount of energy.	3. Release or absorb relatively slight amounts of energy.
4. Rate of reaction is not influenced by external factors.	4. Rate of reaction depends factors such as concentrations, pressure, temperature, and catalyst.

Henri Becquerel discovered natural radioactivity in 1896 coming from a uranium compound. By studying these rays, Ernest Rutherford showed that atoms of one element can be converted into atoms of another element by spontaneous nuclear disintegrations. Many years later it was shown that one element can be transformed into another by nuclear reactions initiated by bombardment of nuclei with accelerated subatomic particles.

Becquerel's discovery led other investigators, including Marie and Pierre Curie, to discover and study new radioactive elements. Radioisotopes are radioactive isotopes.

Fundamental Particles of Matter

PARTICLE	MASS	CHARGE
Electron (e⁻)	0.0005458 amu	1-
Proton (p or p⁺)	1.0073 amu	1+
Neutron (n or n⁰)	1.0087 amu	none

Nuclear **fission** is the splitting of a heavy nucleus into lighter nuclei. Nuclear **fusion** is the combination of light nuclei to make a heavier nucleus. Extremely large amounts of energy are released when these events occur.

26-1   The Nucleus

The principal subnuclear particles are protons and neutrons, with electrons determining the extranuclear structure. Nuclear diameters are about 10^{-4} Angstroms or 10^{-12} cm (atomic diameters are about 1 Angstrom). Thus nuclei are extremely dense with density of approximately 2.44×10^{14} g/cm^3.

Although it would seem that the positively charged protons (and uncharged neutrons) might experience severe electrostatic repulsion crowded into such a small volume, nonradioactive nuclei are stable. It is thought that many of the short-lived subatomic particles (in addition to protons, neutrons and electrons) that have been detected as products of nuclear reactions are involved in binding nuclear particles (**nucleons**) together. These strong forces appear to act only over very short distances, about 10^{-13} cm.

26-2   Neutron-Proton Ratio and Nuclear Stability

The term **nuclide** is used to refer to different atomic forms of *all* elements, whereas the term "isotope" applies only to different forms of the *same* element. Most naturally occurring nuclides have even numbers of protons and even numbers of neutrons. Table 26-2 (reproduced below) shows the abundance of naturally occurring nuclides.

Table 26-2	Abundance of Naturally Occurring Nuclides			
Number of protons	even	even	odd	odd
Number of neutrons	even	odd	even	odd
Number of such nuclides	157	52	50	4

Nuclides with certain "magic numbers" of neutrons and protons are especially stable. Nuclides with a number of nucleons equal to 2, 8, 20, 50, 82, or 126 have unusual stability. This suggests an energy level or shell model for nuclear structure similar to that for electrons in the extranuclear structure.

Example nuclides with magic numbers of nucleons include

_____   _____   _____   _____   _____

A plot of the number of neutrons (N) versus the number of protons (Z) for the stable nuclides is given in Figure 26-1. Above atomic number 20 the most stable nuclides have more neutrons than protons.

26-3   Nuclear Stability and Binding Energy

The **mass deficiency, Δm,** for a nucleus is the difference between the sum of the masses of protons, neutrons and electrons in the atom (the calculated mass) and the actual measured mass of the atom:

$$\Delta m = (\text{sum of masses of all } p^+, n^0, \text{ and } e^-) - (\text{actual mass of the atom})$$

The mass deficiency in the mass that has gone into **nuclear binding energy**. The Einstein mass-energy equivalence relationship, $E = mc^2$, can be rewritten as

$$BE = (\Delta m)c^2$$

Example 26-1:   Calculate the mass deficiency for potassium-39 atoms. The actual mass of a potassium-39 atom is 39.32197 amu.

Example 26-2:   Calculate the nuclear binding energy of ^{39}K in joules/mol of K atoms. 1 joule = 1 kg•m^2/s^2.

Figure 26-2 is a plot of average binding energy expressed as kJ/g of nuclei versus mass number. The nuclei with the highest binding energies (mass numbers 40 to 150) are the most stable.

26-4 Radioactive Decay

Nuclei whose neutron-to-proton ratio lies outside the belt of stability experience spontaneous radioactive decay by emitting one or more particles and/or electromagnetic rays. The kind of decay depends on where the nucleus is positioned relative to the band of stability (Figure 26-1). A summary of the common types of decays is given in Table 26-3. This table also indicates the penetration capabilities of these radiations. The particles can be emitted with different kinetic energies. The energy change is related to the change in binding energy from reactant to products.

26-5 Equations for Nuclear Reactions

Two conservation principles hold for equations for nuclear reactions (the following hold for **all** nuclear reactions)

sum of mass numbers of reactants	=	sum of mass numbers of products
sum of atomic numbers of reactants	=	sum of atomic numbers of products

or for $^{M_1}_{Z_1}Q \rightarrow \, ^{M_2}_{Z_2}R + M_3Z_3\,Y$ (generalized equation)

$$M_1 = M_2 + M_3 \quad \text{and}$$

$$Z_1 = Z_2 + Z_3$$

where the M's are mass numbers and the Z's are atomic numbers.

26-6   Neutron-Rich Nuclei (Above the Band of Stability)

Nuclei in this region have too high a ratio of neutrons to protons. The decays they experience are those that lower the ratio, such as **beta emission**, which is associated with the conversion of a neutron to a proton.

Note that beta emission simultaneously decreases the number of neutrons (by one) and increases the number of protons (by one). Examples of beta emission

$$^{14}_{6}C \rightarrow$$

$$^{226}_{88}Ra \rightarrow$$

26-7 Neutron-Poor Nuclei (Below the Band of Stability)

Nuclei in this region can *increase* their neutron-to-proton ratio by undergoing **electron capture** (K capture), **positron emission**, or **alpha emission**.

Electron capture (K capture) is a process in which an electron from the K shell is captured by the nucleus and a proton is converted to a neutron.

$$^{1}_{1}p + {}^{0}_{-1}e \rightarrow {}^{1}_{0}n$$

$$^{37}_{18}Ar + {}^{0}_{-1}e \rightarrow$$

A positron has the mass of an electron but has a positive charge ($^{0}_{+1}e$). Associated with positron emission is the conversion of a proton into a neutron.

$$^{1}_{1}p \rightarrow {}^{0}_{+1}e + {}^{1}_{0}n$$

$$^{39}_{19}K \rightarrow {}^{0}_{-1}e +$$

$$^{15}_{8}O \rightarrow {}^{0}_{-1}e +$$

Alpha emission occurs for some nuclei, especially heavier ones. Alpha particles are helium nuclei, $^{4}_{2}He$, containing two protons and two neutrons. Alpha emission increases the neutron-to-proton ratio, e.g.,

$$^{204}_{82}Pb \rightarrow {}^{200}_{80}Hg + {}^{4}_{2}He$$

446

26-8   Nuclei With Atomic Number Greater Than 83

All nuclides have atomic numbers greater than 83 are beyond the belt of stability and are radioactive. Many of these isotopes decay by emitting alpha particles.

$$^{238}_{92}U \rightarrow {}^{234}_{90}Th + {}^{4}_{2}He$$

Some of these nuclides also decay by beta emission, positron emission, and electron capture.

The **transuranium elements** (Z≥92) also decay by **nuclear fission** in which the heavy nuclide splits into nuclides of intermediate mass and neutrons.

$$^{252}_{98}Cf \rightarrow {}^{142}_{56}Ba + {}^{106}_{42}Mo + 4\,{}^{1}_{0}n$$

26-9   Detection of Radiations

The particles and radiations emanating during radioactive decay are energetic and some carry charges. The several detection schemes available depend on these facts. The method chosen for a particular experiment depends on which nuclear particles are to be detected.

Photographic Detection

Radiation affects photographic plates or film as does ordinary light. Although the amount of darkening in the photographic negative is related to the amount of radiation, it is difficult to use this approach in a quantitative manner.

Fluorescence Detection

Fluorescent substances can absorb energy from high energy rays and then relax by emitting visible light. A **scintillation counter** is an instrument using this principle.

Cloud Chambers

A cloud chamber contains air saturated with a vapor. Particles emitted in radioactive decay ionize air molecules in the chamber and the vapor subsequently condenses on these ions. Photographing these tracks permits detailed study of their nature.

Gas Ionization Counters

The ions produced by the passage of ionizing radiation between high voltage electrodes causes a current to flow between the electrodes. The current is subsequently amplified. This is the basis of operation of gas ionization counters such as the **Geiger-Mueller counter** (Figure 26-5).

26-10   Rates of Decay and Half-Life

The rates of all radioactive decays are independent of temperature and obey first order kinetics. The rates of decay of different radioactive nuclides vary widely from fractions of a second to millions of years. The same mathematical descriptions of first order decay (decreases in concentration) developed in Section 16-8 apply here.

$$\text{Rate of decay} = k[A] \quad \text{or}$$

$$\ln \frac{A_0}{A} = akt \quad \text{or} \quad \log \frac{A_0}{A} = \frac{akt}{2.303}$$

Where A represents the amount of nuclide remaining after time t, and A_0 is the amount present at the beginning of the observation. The k is the specific rate constant that is characteristic of a particular nuclide. A similar relationship can be used for counting radioactive decays, if N represents the number of disintegrations per unit time.

$$\ln \frac{N_0}{N} = akt \quad \text{or} \quad \log \frac{N_0}{N} = \frac{akt}{2.303}$$

The half-life, $t_{1/2}$, is related to the rate constant by the simple relationship

$$t_{1/2} = \frac{\ln 2}{k} = \frac{0.693}{k}$$

Example 26-3:   How much cobalt-60 remains after 15.0 years? Cobalt-60 has a half-life of 5.27 years.

26-11 Decay Series

Many nuclides are so far away from the belt of stability, that it takes many nuclear disintegrations (a series of them) to attain nuclear stability. Table 26-4 outlines in detail three disintegration series, the ^{238}U, ^{235}U and ^{232}Th series. For any particular decay step, the decaying nuclide is the **parent** nuclide and the product nuclide is the **daughter.**

26-12 Uses of Radionuclides

Practical applications of radionuclides depend on either the use of known decay rates or the continuous emission of radiation.

Radioactive Dating

Radiocarbon dating can be used to estimate the ages of items of organic origin. Carbon-14 is produced continuously in the upper atmosphere by the bombardment of nitrogen-14 by cosmic-ray neutrons:

$$^{14}_{7}N + ^{1}_{0}n \rightarrow ^{14}_{6}C + ^{1}_{1}H$$

The ^{14}C atoms react with O_2 to form CO_2. The CO_2 then is incorporated into plant life by photosynthesis. Living matter has the same ^{14}C fraction as the atmosphere. When the material is no longer living (and is perhaps transformed into some artifact) the ^{14}C fraction decreases because of the radioactive decay of carbon-14. The half-life of ^{14}C is 5730 years.

$$^{14}_{6}C \rightarrow ^{14}_{7}N + ^{0}_{-1}\beta$$

The potassium-argon and uranium-lead methods are used for dating older objects.

$$^{40}_{19}\text{K} \rightarrow \qquad\qquad\qquad t_{1/2} =$$

$$^{238}_{92}\text{U} \rightarrow \qquad\qquad\qquad t_{1/2} =$$

Example 26-4:   Estimate the age of an object whose ^{14}C activity is only 55% that of living wood.

Medical Uses

Cobalt radiation treatment is used for cancerous tumors. Many other nuclides are used in medicine as *radioactive tracers*. A radiation detector can be used to follow the pat of the element throughout the body.

Positron emission tomography (PET) is a form of imaging that uses positron emitters. Isotopes used in this technique are short-lived positron emitters such at ^{11}C, ^{13}N, ^{15}O, and ^{18}F, The patient is placed into a cylindrical gamma ray detector. When the radioisotopes decay, the emitted positron quickly encounters and electron and reacts in a matter-antimatter annihilation, to give off two gamma rays in opposite directions as shown below.

Agricultural Uses

Gamma irradiation of some foods allows them to be stored for longer periods without spoiling.

Industrial Uses

When strips or sheets of metal are manufactured at a definite thickness, the penetrating powers of various kinds of radioactive emissions are utilized. The thickness of the metal is determined by the intensity of radiation passing through it.

Research Applications

Chemical reaction pathways can by investigated using radioactive tracers.

26-13   Artificial Transmutations of Elements

Bombardment of a nuclide with a nuclear particle can make an unstable compound nucleus that decays to a new nuclide by emission of a different particle. The rules for balancing equations for nuclear reactions still hold.

sum of mass numbers of reactants	=	sum of mass numbers of products
sum of atomic numbers of reactants	=	sum of atomic numbers of products

Bombardment with Positive Ions

When the bombarding particle is positively charged it must be accelerated to sufficient energy to overcome the coulomb repulsion of the positive nucleus so that the bombarding particles penetrate that nucleus. Particle accelerators such as **cyclotrons** or **linear accelerators** are used for this. Example reactions are

Neutron Bombardment

Because neutrons have no charge, there is no coulomb repulsion to their nuclear penetration, so they do not have to be accelerated. However, neutron beams are needed. Nuclear reactors are often used as neutron sources. If the neutrons have high kinetic energy they are called **fast neutrons**. **Slow neutrons** ("thermal neutrons") have had their excess energy decreased by collisions with **moderators** such as hydrogen, deuterium, oxygen, or the carbon atoms in paraffin. Slow neutrons are more likely to be captured by target nuclei. Example reactions are

26-14 Nuclear Fission

Some nuclides with atomic numbers greater than 80 are able to undergo fission in which they are split into nuclei of intermediate masses and emit one or more neutrons. Some of these fissions are spontaneous while others require activation by neutron bombardment. Enormous amounts of energy are released in these fissions. Some of the possible ways that ^{235}U can split (after bombardment by a neutron) are

$$^{235}_{92}U + ^{1}_{0}n \longrightarrow [^{236}_{92}U] \longrightarrow$$

Fission is energetically favorable for elements with Z greater than 80 because the product nuclides are more stable (near the high part of the nuclear binding energy curve).

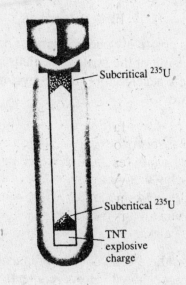

Subcritical ^{235}U

Subcritical ^{235}U

TNT explosive charge

26-15 Nuclear Fission Reactors

In a nuclear fission reactor, the fission reaction is controlled by inserting materials to absorb some of the neutrons so that the mixture does not explode. The energy produced can then be used in a power plant as a heat source.

Light Water Reactors

Light water reactors use normal, albeit highly purified, water as both the coolant and moderator. Often the water is pressurized so that the temperature is higher to improve the turbine efficiency. There are also unpressurized or boiling water reactors. Light water reactors are the most common commercial nuclear reactors for the generation of electricity. There are approximately one hundred of these in operation in the United States.

Fuel

Most commercial nuclear reactors used enriched $^{235}UO_2$ as their fuel. This is a solid that can be compressed into small cylinders. These cylinders are then stacked into a large Zirconium alloyed metal tube called a fuel rod. These fuel rods are then assembled into the reactor to make the core. After the fuel has been consumed, the fuel rods are removed and left in cooling water to allow the fission products to decay. At some point they must be disposed of in a safe and efficient manner. In the United States, the present fuel rod disposal plan is to bury the material in a specially designed mine in Yucca Flats, Nevada at some point early in the 21st Century.

Moderator

Moderators are materials that slow down fast neutrons, turning them into thermal neutrons. These thermal neutrons have a high probability of being captured by another fissile nucleus, such s ^{235}U or ^{239}Pu. This sustains the nuclear chain reaction. Several materials have been used as moderators including graphite, water, and "heavy water". (Heavy water is water that has the two normal 1H atoms replaced with ^2H, or deuterium atoms.) Most American nuclear reactors used water as the moderator. Some Canadian reactors use heavy water. The Chernobyl reactor was graphite moderated.

Control Rods

The preferred method to control nuclear reactors is to absorb the neutrons before they can cause a fission reaction. In effect, this interrupts the self-sustaining chain reaction and causes the fission rate to decrease. Long rods of neutron absorbing material, primarily elemental boron, are very effective control rods. When it is time to start reactor operation, the control rods are removed from the reactor core. To stop the reactor, the rods are inserted into the core. To maintain a stable reactor during operation, the rods are continuously inserted and removed by small fractions of an inch. Computers control this phase of operation and keep the reactor stable.

Cooling Systems

To keep reactors from melting down, they must be cooled at all times while they are in operation, and for some time period after they have been shut down. In most commercial and military nuclear reactors the coolant of choice is water. Not only does water have a lot of critical factors that play in its favor for a coolant, such as a large heat capacity, but it is also cheap and easy to use. For many reactors, water is both the coolant and the moderator. Other coolants that have been used in some reactors include gases, primarily He, and liquid sodium.

However, if the cooling water is disrupted there is no way for the core to cool and melting can occur. In the reactor accident at Three Mile Island, core-cooling water was stopped and the reactor began to meltdown. However, the reactor operators realized their error and managed to get cooling water back on the reactor before a complete meltdown occurred. In the case of the Chernobyl reactor accident, the causes are more complicated than simply cutting off the water supply to the core but the result was the same. In this case, not only did the core meltdown but a steam/hydrogen explosion occurred that blew the reactor's lid off. This exposed the reactor core to the atmosphere and precipitated the world's largest nuclear reactor accident.

Shielding

Because fission reactors produce enormous numbers of neutrons as well as nuclear particles from the radioactive decays in the fission products, humans must be shielded from the potentially lethal radiation. Many materials are used for shielding but lead and concrete are the two most common. Commercial reactors primarily use concrete because they must have large containment vessels over the reactor. These containment buildings are usually steel reinforced concrete providing both shielding and structural integrity. Nuclear reactors, particularly those on submarines, use lead because it saves space onboard the ship.

Breeder Reactors

Breeder reactors manufacture more fuel than they used. It will generate electrical power as well as maximize neutron capture in the core by $^{238}_{92}U$.

Nuclear Power: Hazards and Benefits

The enormous amounts of energy released by fissioning atoms can be used in several possible ways. One is to generate electricity, which is the method employed in nuclear power reactors used in commercial and military reactors. Nuclear energy plants do not pollute the air with oxides of sulfur, nitrogen, carbon, and particulate matter. The containers that the long – lived radionuclides are kept in are stored underground and could corrode over long periods of time.

Nuclear Weapons

Another use is much more destructive and that is nuclear weapons. When fission was initially discovered by Otto Hahn and Fritz Strassmann in 1939 and explained by Otto Frisch and Lise Meitner less than a month later, Europe was on the brink of World War II. Several notable exiled scientists, including Leo Szilard and Eugene Wigner, realized that the German scientists had stumbled upon a discovery that theoretically could be converted into a devastating weapon. All that was required to complete the process was a method to make the nuclear fission into a **self-propagating chain reaction**. Earlier in 1933, Szilard had theoretically determined a method to sustain these chain reactions. With Hahn and Strassmann's publication, the pieces of the puzzle began to fall together for the exiled scientists. For these scientists, who had recently escaped the Nazi regime, the thought that Germany could build nuclear weapons was abhorrent. Thus they convinced President Franklin Roosevelt to begin a hurried weapons development program that became known as the Manhattan Project. By late August of 1945, under the direction of Robert Oppenheimer, the United States had built and exploded three nuclear fission weapons. The first was a test explosion in New Mexico and the other two were dropped on Hiroshima and Nagasaki, Japan.

These first nuclear weapons were relatively simple. A diagram of the type of weapon dropped on Hiroshima is given below. In this device, two small pieces of highly **enriched** ^{235}U are compressed into a single piece of U that exceeds the critical mass, the amount of U necessary to sustain the chain reaction. When the two pieces are compressed, the nuclear fusion reaction begins until the energy output is so great that the bomb explodes. The fission products are mixed with the dust and debris from the explosion to form radioactive **fallout**.

26-16 Nuclear Fusion Energy

At the present time, it appears that the most energetic processes in the universe are nuclear fusion reactions. These are the fusing or merging of two light nuclei into a heavier nucleus. In fact, it is this energy source that fuels the stars and ultimately creates all of the chemical elements. If humans could harness this energy source, it would potentially provide all of the human energy requirements for the remainder of the earth's lifetime. We have already learned how to harness fusion for destructive purposes.

Fusion, the merging of light nuclei elements to make heavier nuclei, is favorable for the very light atoms. Extremely high energies or temperatures are necessary to initiate fusion reactions.

Thermonuclear, or hydrogen, bombs are nuclear fusion reactions that are unleashed in a few fractions of a second. In effect, detonating a hydrogen bomb is the equivalent of igniting a star in your enemy's backyard. The effect is predictable. Enormous amounts of heat and energy literally vaporize the surrounding area for several square miles. Luckily, none of these devices have been used on human populations. If we could take that same energy and find a way to harness it, humans would benefit from the technology rather than fear it.

Nuclear physicists and chemists have tried for years to develop nuclear fusion reactors that are analogous to fission reactors. However, there is one enormous hurdle to overcome in fusion reactor development. Fusion takes place at temperatures of 10 million degrees centigrade! In other words we have to design a structure that can withstand that temperature and not melt. It is a testament to the ingenuity of scientists that there are a few fusion reactors at various places around the world. None of these reactors are at the stage of producing electricity for commercial use. However, over the last twenty years, the temperatures and reaction conditions inside these reactors have inched closer to the point that a sustained fusion reaction is possible. It would appear that right at the start of the 21st Century, humans are on the verge of using the fuel of the stars to supply their terrestrial needs.

Synthesis Question

How are thermonuclear of fusion reactors designed so that the hot plasma, temperature of approximately 10 million degrees, does not touch the sides of the reactor? The reactor would melt if the plasma were to touch the sides.

Group Question

Stars are enormous thermonuclear fusion reactors generating enormous amounts of heat and energy. What keeps stars from blowing themselves apart? How do they remain stable for millions and billions of years?

Chapter Twenty Seven

ORGANIC CHEMISTRY I: FORMULAS, NAMES AND PROPERTIES

All living matter contains many compounds of carbon. Carbon atoms bond to each other (catenation, "chain-making") to a much greater extent than atoms of any other element. This accounts for the enormous number of carbon-containing compounds. Most compounds of carbon contain chains and rings formed by carbon atoms bonding to each other. Carbon atoms also bond to many other atoms, such as H, N, O, S, P, and the halogens. Most organic compounds are covalent - organic chemistry may be described as the study of the covalent compounds of carbon.

SATURATED HYDROCARBONS

27-1 Alkanes and Cycloalkanes

Hydrocarbons contain only carbon and hydrogen. Saturated hydrocarbons contain only single bonds, i.e., only sigma (σ) bonds. Petroleum and natural gas are the main sources of saturated hydrocarbons.

Alkanes

The alkanes are the simplest saturated hydrocarbons (Figure 27-1). Methane, CH_4, is the first member of this homologous series, a family of compounds in which each member differs from the next by a specific number and kind of atoms. Each alkane differs from the next by a methylene group, CH_2.

The bonding in methane was described in Section 8-7. Review that section. See also Figure 27-2. Each C atom is sp^3 hybridized in the alkanes, i.e., each C atom forms four σ bonds.

The general formula for the alkanes is C_nH_{2n+2}. Ethane, C_2H_6, and propane, C_3H_6, are the next two members of the family (Figures 27-3 and 27-4). They consist of interlocking tetrahedra.

Isomers are substances that have the same molecular formulas, but different structures.

Two alkanes have the formulas C_4H_{10}. They are **structural isomers**. See Figure 27-5.

Three alkanes have formulas C_5H_{12}, so there are three isomeric pentanes. See Example 27-1 in text and associated margin figures.

There are five isomeric hexanes, C_6H_{14} (Table 27-3). The number of structural isomers increases rapidly with increasing numbers of C atoms (Table 27-4).

In Section 13-9, we described variations in boiling points of similar compounds. When there are no complicating factors such as hydrogen bonding, boiling points of a series of similar compounds increase with increasing molecular weight. Boiling points of alkanes increase with molecular weight (Figure 27-6 and Table 27-2).

Cycloalkanes

Cyclic saturated hydrocarbons are called *cycloalkanes*. They have the general formula C_nH_{2n}. A few examples are

461

27-2 Naming Saturated Hydrocarbons

The IUPAC names for the first 20 "straight-chain" or "normal" alkanes are given in Table 27-2. Other compounds are named as *derivatives* of alkanes.

Branched-chain alkanes are named by the following rules.

1. Choose the longest continuous chain of C atoms - this gives the basic name (stem).

2. Number each C atom in the basic chain, starting at the end that gives the lowest number to the first substituent (group attached to chain).

3. For each substituent on the chain, we indicate *position* (by an Arabic numeric prefix) **and** the kind of substituent (by its name). The position of a substituent on the chain is indicated by the lowest number possible - it precedes the name of the substituent.

4. When there are two or more substituents of a given kind, use prefixes to indicate the number of substituents: di = 2, tri = 3, tetra = 4, penta = 5, hexa = 6, hepta = 7, octa = 8, and so on.

5. The combined substituent numbers and names serve as a prefix for the basic hydrocarbon name.

6. Separate numbers from numbers by commas and numbers from words by hyphens. Words are "run together".

Alkyl groups (represented by R) are common substituents. They are thought of as fragments of alkanes in which one H atom has been removed (general formula C_nH_{2n+1}). In alkyl groups the **-ane** suffix in the name of the parent alkane is replaced by **-yl**. See Table 27-5.

UNSATURATED HYDROCARBONS

The three classes of unsaturated hydrocarbons are the

1. alkenes and cycloalkenes
   noncyclic alkenes containing one double bond have the general
   formula C_nH_{2n}

2. alkynes and cycloalkynes
   noncyclic alkynes containing one triple bond have the general
   formula C_nH_{2n-2}

3. aromatic hydrocarbons

27-3   Alkenes

The simplest alkenes contain one C=C bond per molecule. The general formula for the simplest alkenes is C_nH_{2n}. The first two alkenes are ethene, C_2H_4, (bonding described in Section 8-13) and propene, C_3H_6.

Each doubly bonded C atom is sp^2 hybridized, i.e., forming two σ bonds and one π bond.

Systematic names for alkenes use the same stems as alkanes with the same number of C atoms. The **-ane** suffix is changed to **-ene**. Their common (trivial) names have the same stem, but the suffix **-ylene** is used. In chains of four or more C atoms, a numerical prefix shows the position of the lowest-numbered doubly bonded C atom. Always choose the longest chain that contains the C=C bond.

Polyenes contain two ore more double bonds per molecule. Suffixes, **-adiene**, **-atriene**, and so on, indicate the number of double bonds in polyenes.

Positions of substituents are indicated as they are for alkanes. In naming branched chain alkenes, the C=C bond takes positional preference over substituents, i.e, the position of the C=C bond is given the lowest number possible.

Cycloalkenes containing one double bond have the general formula C_nH_{2n-2}. Some examples are

27-4   The Alkynes

Alkynes contain C≡C bonds. The simplest alkyne is C_2H_2, ethyne, or acetylene (Figure 27-10).

Alkynes with only one C≡C bond have the formula C_nH_{2n-2}. See Section 8-14 for a description of the bonding in acetylene. Each C atom in a C≡C bond is sp hybridized, i.e., forming one σ bond and two π bonds.

Alkynes are named like the alkenes except that the suffix **-yne** is used with the characteristic stem derived from the name of the alkane with the same number of C atoms.

AROMATIC HYDROCARBONS

Historically, the word aromatic was used to describe pleasant smelling substances. Now it refers to benzene, C_6H_6, the derivatives of benzene, and other compounds that have similar chemical properties.

27-5   Benzene

The structure of benzene is (See Figure 27-11):

Benzene contains six sp^2 hybridized carbon atoms. All six C atoms and the six H atoms lie in a plane. The π electrons in benzene belong to all of the C atoms forming a *delocalized* ring structure. Benzene is a very stable molecule due to this electron delocalization.

27-6   Other Aromatic Hydrocarbons

Many aromatic hydrocarbons contain alkyl groups attached to benzene rings (as well as to other aromatic rings). Positions of substituents on benzene rings are indicated by the prefixes, *ortho-* (*o-*) for substituents on adjacent (1,2) C atoms, *meta-* (*m-*) for substituents on C atoms 1 and 3, and *para-* (*p*) for substituents on C atoms 1 and 4.

Coal tar is the common source of benzene and many other aromatic compounds. See Figure 27-12 and Table 27-6.

Some aromatic hydrocarbons that contain fused rings are

Many important naturally occurring compounds contain fused rings, some aromatic and some aliphatic.

27-7   Hydrocarbons: A Summary

Hydrocarbons are organic compounds that contain only carbon and hydrogen. A summary of their classification scheme is given in Figure 27-13.

FUNCTIONAL GROUPS

Functional groups are groups of atoms that represent potential reaction sites. Compounds that contain a given functional group usually undergo similar reactions. Functional groups influence physical properties as well.

27-8   Organic Halides

A halogen atom may replace almost any hydrogen atom in a hydrocarbon. Thus, the functional group is the halide (-X) group. Examples include chloroform, $CHCl_3$; 1,2-dichloroethane, $ClCH_2CH_2Cl$ and para-dichlorobenzene

27-9   Alcohols and Phenols

The functional group in alcohols and phenols is the hydroxyl (-OH) group. Alcohols and phenols can be considered derivatives of hydrocarbons in which one or more H atoms have been replaced by -OH groups. Perhaps the better view is that alcohols are derivatives of water (Section 8-9) in which one H has been replaced by an aliphatic group (R). Phenols are derivatives of water in which one H has been replaced by an aromatic group (aryl group, Ar).

H-O-H CH_3OH

water   methyl alcohol   phenol

Ethyl alcohol (ethanol) is the most familiar alcohol. Phenol is the most familiar phenol. See Figures 27-14 and 27-15.

Alcohols are considered neutral compounds because they are only *very* slightly acidic. Phenols are weakly acidic. $K_a = 1.0 \times 10^{-10}$ for phenol, which is a very weakly acidic, but a very corrosive compound.

We classify alcohols into three classes to help distinguish their reactivity.

a primary (1°) alcohol   a secondary (2°) alcohol   a tertiary (3°) alcohol

466

Nomenclature

The stem for the parent hydrocarbon *plus* an **-ol** suffix is the systematic name for an alcohol. A numeric prefix indicates the position of the -OH group in alcohols with three or more C atoms. Common names are the name of the appropriate alkyl group *plus* alcohol.

There are several isomeric monohydric acylic (contains no rings) alcohols that contain more than three C atoms. There are four isomeric four-carbon alcohols.

There are eight isomeric five-carbon alcohols.

Polyhydric alcohols contain more than one -OH group per molecule.

Phenols are usually called by their common (trivial) names.

Physical Properties

Because the -OH group is quite polar, whereas alkyl groups are nonpolar, the properties of alcohols depend upon (1) the number of -OH groups per molecule and (2) the size of the organic group. Boiling points of monohydric (one -OH per molecule) alcohols increase, while their solubilities in water decrease with increasing molecular weight. See Table 27-8.

Polyhydric alcohols are more soluble in water because they contain two or more polar groups (-OH). Most phenols are solids at room temperature. Most are only slightly soluble in water, unless they contain other polar groups.

27-10   Ethers

Ethers may be thought of as derivatives of water in which both H atoms have been replaced by alkyl or aryl groups.

H-O-H

water

CH$_3$-O-H

an alcohol

CH$_3$-O-CH$_3$

an ether

Ethers are not very polar and not very reactive. They are excellent solvents. Common names are used for most ethers.

27-11   Aldehydes and Ketones

The functional group in aldehydes and ketones is $-\overset{\overset{\displaystyle O}{\|}}{C}-$, the carbonyl group. Except for formaldehyde, aldehydes have one H atom and one organic group bonded to a carbonyl group. Ketones have two organic groups bonded to a carbonyl group. See Figure 27-16. The organic groups may be aliphatic or aromatic; or one of each.

$\overset{\overset{\displaystyle O}{\|}}{CH_3-C-H}$

aldehydes

$\overset{\overset{\displaystyle O}{\|}}{CH_3-C-CH_3}$

ketones

Common names for aldehydes are derived from the name of the acid with the same number of C atoms. IUPAC names are derived from the parent hydrocarbon name by replacing -*e* with **-al**.

Ketones are often called by their common names. These are usually just the names of the alkyl or aryl groups attached to the carbonyl group, followed by "ketone".

The IUPAC name for a ketone is the characteristic stem for the parent hydrocarbon plus the suffix **-one** (pronounced "own"). A numeric prefix indicates the position of the carbonyl group in a chain or on a ring.

Many aldehydes and ketones occur in nature.

27-12 Amines

Amines are derivatives of ammonia in which one or more H atoms have been replaced by organic groups (aliphatic or aromatic or a mixture of both).

Structure and Nomenclature

See bottom of page 1001. There are three classes of amines.

ammonia   1° amines   2° amines   3° amines

Like NH_3, the amines are basic because the N atom has a lone pair of electrons. Low molecular weight aliphatic amines are usually stronger bases than NH_3. See Table 28-2.

Heterocylic amines have one or more N atoms *in* a ring structure. Many are important in living systems.

27-13 Carboxylic Acids

Carboxylic acids contain the carboxyl group, $-\overset{\overset{\displaystyle O}{\|}}{C}-O-H$, as their functional group.

Their general formula is $R-\overset{\overset{\displaystyle O}{\|}}{C}-O-H$, where R is an alkyl *or* an aryl group. See Figure 27-17.

In the IUPAC name for a carboxylic acid, *e* is dropped from the name of the parent hydrocarbon and the suffix **-oic** is added, followed by the word **acid**. See Table 27-11. Many organic acids are called by their common (trivial) names which are derived from Greek or Latin. (Tables 27-11 and 12).

Positions of substituents on chains in carboxylic acids are indicated by numeric prefixes as in other compounds (counting from the carboxylic C atom). They are also often indicated by lower case Greek letters.

27-14   Some Derivatives of Carboxylic Acids

Four important classes of compound contain acyl groups and are considered derivatives of carboxylic acid. The following R's may represent alkyl *and* aryl groups.

acyl halides   **esters**   **amides**
(acid halides)

Acyl Halides (Acid Halides)

Acyl halides are much more reactive, and more volatile, than their parent acids. They react with water to form their parent acids and a hydrohalic acid.

Esters

Heating a carboxylic acid with an alcohol in the presence of a small amount of an inorganic acid, such as H_2SO_4, produces an equilibrium mixture. It contains some ester and water, as well as unreacted acid and alcohol. See Figure 27-18.

Esters are usually called by their common names. See Table 27-13. Many simple esters have pleasant odors and occur naturally. Esters are used in fragrances and as artificial flavors. Low molecular weight esters are excellent solvents.

Fats are *solid* esters of glycerol and most saturated acids. **Oils** are *liquid* esters of glycerol and mostly unsaturated acids. The "acid" parts of fats and oils usually contain even numbers of C atoms (16 and 18 are most common). See Figure 27-19.

Some acids that are found (as their esters) in fats and oils include:

Naturally occurring fats and oils are mixtures of esters. Two common examples are

Triglycerides are the triesters of glycerol. The common name for triglycerides is **tri(stem for the acid part)** plus an **-in** suffix.

Waxes are esters of long chain fatty acids and alcohols other than glycerol. Most are derived from monohydric alcohols. Beeswax and carnauba wax are esters of myricyl alcohol, $C_{30}H_{61}OH$.

Amides

Amides are derivatives of organic acids and primary or secondary amines. The functional group of amides are:

Amides are named as derivative of carboxylic acids. The suffix *-amide* is substituted for *-ic acid* or *-oic acid*.

When an aryl or alkyl substituent is present on the N atom, the letter *N* and the name of the substituent are prefixed to the name of the unsubstituted amide.

27-15   Summary of Functional Groups

A summary based on Figure 27-20 follows on the next page.

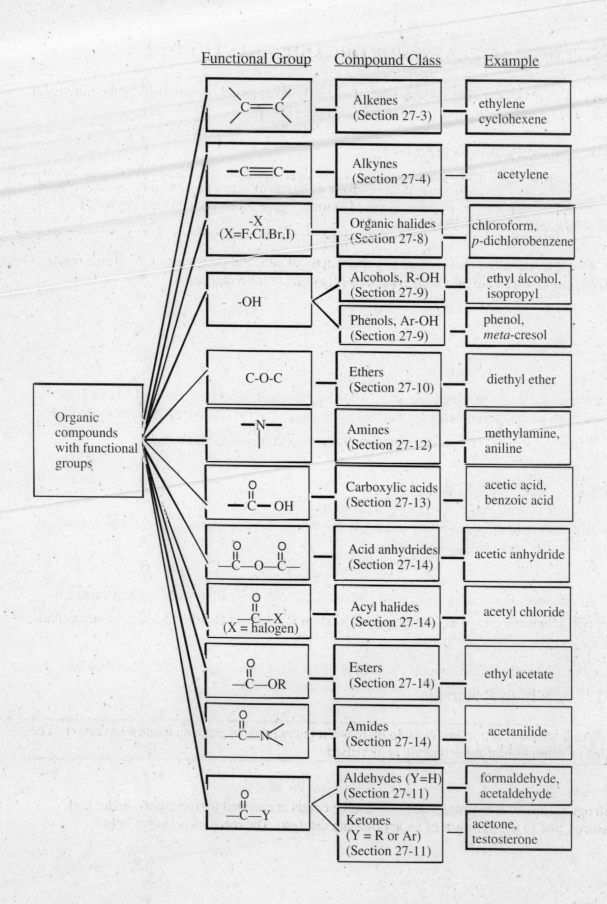

Functional Group	Compound Class	Example
C=C	Alkenes (Section 27-3)	ethylene cyclohexene
—C≡C—	Alkynes (Section 27-4)	acetylene
-X (X=F,Cl,Br,I)	Organic halides (Section 27-8)	chloroform, *p*-dichlorobenzene
-OH	Alcohols, R-OH (Section 27-9)	ethyl alcohol, isopropyl
	Phenols, Ar-OH (Section 27-9)	phenol, *meta*-cresol
C-O-C	Ethers (Section 27-10)	diethyl ether
—N—	Amines (Section 27-12)	methylamine, aniline
—C—OH	Carboxylic acids (Section 27-13)	acetic acid, benzoic acid
—C—O—C—	Acid anhydrides (Section 27-14)	acetic anhydride
—C—X (X = halogen)	Acyl halides (Section 27-14)	acetyl chloride
—C—OR	Esters (Section 27-14)	ethyl acetate
—C—N	Amides (Section 27-14)	acetanilide
—C—Y	Aldehydes (Y=H) (Section 27-11)	formaldehyde, acetaldehyde
	Ketones (Y = R or Ar) (Section 27-11)	acetone, testosterone

Organic compounds with functional groups

472

FUNDAMENTAL CLASSES OF ORGANIC REACTIONS

Reactivity depends on structure. The reactivity is largely determined by the functional groups and by the structure *near* the active functional groups.

27-16 Substitution Reactions

In a substitution reaction an atom or group of atoms attached to a carbon atom is replaced (substituted for) by another atom or group of atoms. There is no change in the degree of saturation at the reactive carbon atom.

Halogenation reactions are an important class of substitution reactions. Chlorine reacts with alkanes in **free radical chain reactions** (also substitution reactions).

Many substitution reaction of alkanes produce more than one product.

Nitration reaction of an aromatic hydrocarbon replaces an H atom attached to an aromatic ring with a nitro, $-NO_2$, group.

27-17 Addition Reactions

An addition reaction involves an increase in the number of groups attached to carbon. The degree of saturation of the molecule is increased.

Hydrogenation adds hydrogen across a double bonds at elevated temperatures, under high pressures, and in the presence of an appropriate catalyst. The reaction is shown below.

The **hydration reaction** is the addition of water to an alkene. The reaction where ethanol is produced is

27-18 Elimination Reactions

An **elimination** reaction involves the removal of groups attached to carbon. The degree of unsaturation increases.

Dehydrohalogenation is an important kind of elimination reaction.

27-19 Polymerization Reactions

A **polymer** is a large molecule that is typically a high-molecular weight chain of small molecules. The small molecules that have been joined to form the polymer are called **monomers**. Polymerization is the combination of many small molecules (monomers) to form large molecules (polymers).

Addition polymerization

$$n \ CH_2=CH_2 \xrightarrow{\text{catalyst}}$$

$$n \ CF_2=CF_2 \xrightarrow{\text{catalyst}}$$

<u>Vulcanization</u>

<u>Formation of rubber</u>

See Table 27-14 for a list of important addition polymers.

If two different monomers are mixed and then polymerized, **copolymers** are formed.

Condensation polymerization

   Condensation polymers occur when two molecules react and eliminate a small molecule. Molecules eliminated commonly are water and HCl. Important condensation polymers include nylon, dacron, and kevlar.

   Dacron is used in clothing to make it wrinkle free. Another unusual but useful property of dacron is that blood does not clot in contact with dacron. Consequently, it is used to make artificial arteries.

Nylons are widely used in a variety of commercial products including stockings, rope, guitar strings, and fire-proof clothing.

Synthesis Question

TNT, the explosive ingredient in dynamite, has the correct name of 2,4,6-trinitrotoluene. Draw the structure of TNT.

Group Question

Aerobic respiration produces carbon dioxide and water as its end products. Anaerobic respiration has different end products. What are the end products of anaerobic respiration? How could you easily detect that someone has switched from aerobic to anaerobic respiration?

Chapter Twenty Eight

ORGANIC CHEMISTRY II: SHAPES, SELECTED REACTIONS, AND BIOPOLYMERS

SHAPES OF ORGANIC MOLECULES

Isomers are substances that have the same molecular formulas, but different structures.

28-1 Constitutional Isomers

Two alkanes have the formula C_4H_{10}. They are **constitutional (or structural) isomers** (differ in the order in which their atoms are bonded together). See Figure 27-5. Three alkanes have the formula C_5H_{12} so there are three isomeric pentanes. See Figures on page 1034.

There are five isomeric hexanes, C_6H_{14} (Table 27-3). The number of structural isomers increases rapidly with increasing numbers of C atoms (Table 27-4).

28-2   Stereoisomers

In stereoisomers, although the atoms are linked in the same atom-to-atom order, their spatial arrangements are different.

Geometrical Isomerism

Geometrical isomers differ only in the spatial orientation of groups about a plane or direction. See Figures 28-1 and 28-2.

cis-1,2-dichloroethene   *trans*-1,2-dichloroethene

Optical Isomerism

An *asymmetric* carbon atom has four different atoms, or groups of atoms, attached to it. A compound that contains an asymmetric C atom can exist in two forms that are *mirror images* of each other, i.e., they are *not* superimposable on each other. See Figures 28-3 and 28-4.

The two mirror image forms (**optical isomers**) of a compound are called **enantiomers**. They rotate plane-polarized light by the same amount in opposite directions. See Figures 25-4 and 25-5.

The dextrarotatory (d-) isomer of a compound rotates plane polarized light to the right. The levorotatory (l-) isomer rotates plane polarized light to the left by the same amount. A racemic mixture is a single sample containing equal amounts of the two optical isomers of a compound.

28-3 Conformations

A conformation is *one specific geometry* of a molecule. Conformations of a compound differ by the extent of rotation about a single bond (Figures 28-7, 28-8 and 28-9).

ethane
staggered conformation

ethane
eclipsed conformation

cyclohexane
chair conformation

cyclohexane
boat conformation

SELECTED REACTIONS

28-4   Reactions of Brønsted-Lowry Acids and Bases

Many organic compounds can act as weak Brønsted-Lowry acids or bases. Such reactions involve the transfer of H^+ ions or protons (Section 10-4). These acid-base reactions, like similar reactions of inorganic acids and bases, are usually fast and reversible. Therefore, these reactions may be discussed in terms of equilibrium constants (Section 18-4).

Some Organic Acids

$$\underset{acid_1}{CH_3COOH} + \underset{base_2}{H_2O} \rightleftharpoons \underset{acid_2}{CH_3COO^-} + \underset{base_1}{H_3O^+} \qquad K_a = \frac{[CH_3COO^-][H_3O]}{[CH_3COOH]} = 1.8 \times 10^{-5}$$

We can rank the acid strengths of some common organic species

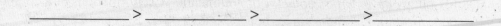

Some Organic Bases

		base$_1$	acid$_2$		acid$_1$	base$_2$		

primary, 1° RNH_2 + H_2O $\rightleftharpoons$ RNH_3^+ + OH^- $K_b = \dfrac{[RNH_3^+][OH^-]}{[RNH_2]}$
amine

secondary, 2° R_2NH + H_2O $\rightleftharpoons$ $R_2NH_2^+$ + OH^- $K_b = \dfrac{[RNH_3^+][OH^-]}{[RNH_2]}$
amine

tertiary, 3° R_3NH + H_2O $\rightleftharpoons$ R_3NH^+ + OH^- $K_b = \dfrac{[R_3NH^+][OH^-]}{[R_3N]}$
amine

We can rank the base strengths of some common organic species

_____ > _____ ≈ _____ > _____ ≈ _____ ≈ _____

28-5   Oxidation-Reduction Reactions

In organic reactions oxidation still refers to the loss of electrons and reduction the gain of electrons. However, it is exhibited slightly differently. **Oxidation** of an organic molecule usually corresponds to *increasing* its *oxygen* content or *decreasing* its *hydrogen* content. **Reduction** of an organic molecule usually corresponds to *decreasing* its *oxygen* content or *increasing* its *hydrogen* content. Examples

$$
\begin{array}{ccc}
H{-}\underset{\underset{H}{|}}{\overset{\overset{H}{|}}{C}}{-}H & \xrightarrow{\text{oxidation}} & H{-}\underset{\underset{H}{|}}{\overset{\overset{H}{|}}{C}}{-}OH
\end{array}
$$

aldehyde

$$
\underset{R}{\overset{O}{\overset{\|}{C}}}{\diagdown}_{H} \xrightarrow{\text{oxidation}}
$$

Oxidation of Alcohol

Aldehydes can be prepared by the oxidation of primary alcohols

1° alcohol

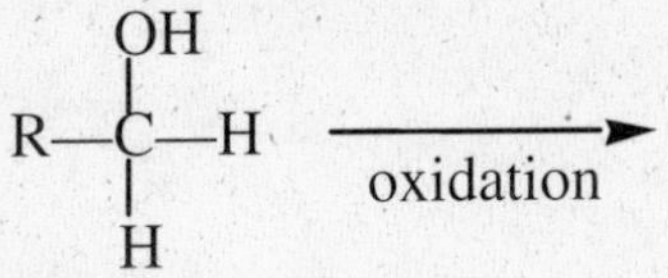

$$R-\overset{\overset{\displaystyle OH}{|}}{\underset{\underset{\displaystyle H}{|}}{C}}-H \xrightarrow{\text{oxidation}}$$

Ketones can be prepared by the oxidation of secondary alcohols

2° alcohol

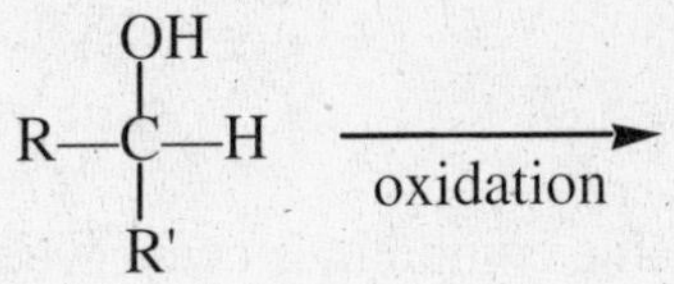

$$R-\overset{\overset{\displaystyle OH}{|}}{\underset{\underset{\displaystyle R'}{|}}{C}}-H \xrightarrow{\text{oxidation}}$$

Oxidation of tertiary alcohols is difficult because the breaking of a carbon-carbon bond is required.

Reduction of Carbonyl Compounds

carboxylic acid

$$R-\overset{\overset{\displaystyle O}{\|}}{C}-OH \xrightarrow{\text{reduction}}$$

ester

$$R-\overset{\overset{\displaystyle O}{\|}}{C}-OR' \xrightarrow{\text{reduction}}$$

aldehyde

$$R-\overset{\overset{\displaystyle O}{\|}}{C}-H \xrightarrow{\text{reduction}}$$

ketone

$$R-\overset{\overset{\displaystyle O}{\|}}{C}-R' \xrightarrow{\text{reduction}}$$

Oxidation of Alkylbenzenes

Aromatic ring systems are quite resistant to oxidation but their alkyl side chains can be oxidized.

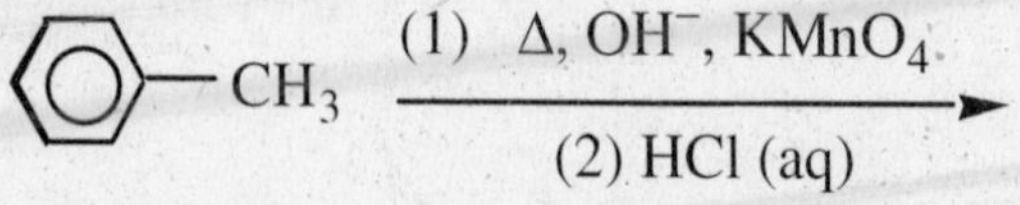

$$\text{(phenyl)}-CH_3 \xrightarrow[\text{(2) HCl (aq)}]{\text{(1) } \Delta,\ OH^-,\ KMnO_4}$$

Combustion of Organic Compounds

The most extreme oxidation reactions of organic compounds occur when they burn in O_2. These combustion reactions are highly exothermic.

$$\underline{\quad} C_8H_{18}\ +\ \underline{\quad} O_2 \longrightarrow$$

28-6   Formation of Carboxylic Acid Derivatives

Formation of acyl halides

$$R-\overset{\overset{\displaystyle O}{\|}}{C}-H\ +\ PCl_5 \longrightarrow$$

Formation of esters

$$R-\overset{\overset{\displaystyle O}{\|}}{C}-OH\ +\ CH_3CH_2OH \underset{}{\overset{H^+,\ \Delta}{\rightleftharpoons}}$$

484

Acyl halides react with alcohols to produce esters much more rapidly than do their parent acids.

$$R-\overset{\overset{\displaystyle O}{\|}}{C}-Cl \quad + \quad CH_3CH_2OH \longrightarrow$$

Formation of amides

$$R-\overset{\overset{\displaystyle O}{\|}}{C}-Cl \quad + \quad CH_3CH_2NH_2 \longrightarrow$$

28-7   Hydrolysis of Esters

Esters are not very reactive. They can be hydrolyzed by refluxing with strong bases. This kind of reaction is called **saponification** (soap-making). The hydrolysis of fats and oils [esters of glycerol (glycerine) and long chain aliphatic acids] produces soaps.

$$R_1-\overset{\overset{\displaystyle O}{\|}}{C}-OR_2 \quad + \quad Na^+OH^- \overset{\Delta}{\longrightarrow}$$

BIOPOLYMERS

Many of the chemicals that are important to living organisms are polymeric in nature. Biopolymers can be categorized into three large classes of molecules: carbohydrates, proteins, and nuclei acids. These molecules are some of the largest molecules in nature. For example, DNA, a nucleic acid, has a molar mass of tens of billions of grams per mole. Their structures are relatively complex but are built around the basic shapes that we learned in Chapter 8. In fact, these molecules are the perfect merger of the basic VSEPR shapes and the organic chemistry introduced in Chapter 27.

28-8   Carbohydrates

The word carbohydrates means hydrated or watered carbons. This is because the general formula for these molecules is $C_n(H_2O)_m$. Thus to early researchers these molecules appeared to be C atoms surrounded by water molecules. However, we now know that these molecules do not have water molecules in them but they do have numerous H atoms and OH groups attached to C atoms. Carbohydrates are an important part of our diets where we eat them as sugars and starches. They are the primary source of energy in our diets.

Monosaccharides (single sugars) are the simplest form of a carbohydrate. Monosaccharides are classified based on two characteristics: the length of the chain and the functional group present in the monosaccharide. For example, ribulose is a five carbon sugar, or a pentose, whereas glucose is a six carbon sugar, or a hexose. Ribulose also contains a ketone functional group and can be called a ketose. Glucose contains an aldehyde functional group and thus may be classified as an aldose.

To condense this naming we could call ribulose a ketopentose and glucose an aldohexose.

Example 28-1:   Fructose is a ketohexose. Draw an acceptable structure of a ketohexose.

See Table 28-4 for the structure of several common monosaccharides. The structure of these sugars are shown in their linear form. However, monosaccharides can form a ring or cyclic form that is an important part of their structure and function. The ring can also exist in two forms, α and β.

Glucose, a very common monosaccharide, is the sugar that we wll have in our blood stream. It is a very important energy source in metabolism and other biological activities. Draw the ring formation steps for glucose. Note how the ring is initially formed and the difference between the α and β forms.

Two monosaccharide units can link together through an elimination reaction to form a new bond called a **glycosidic** linkage. The linkages are designated based on the numbers of the two atoms that are bound together and whether or not the bond is in the α or β position.

Disaccharides are sugars made from two monosaccharides. Maltose is an example of a disaccharide that contains 2 glucose units linked by an α -1,4-glycosidic linkage. Maltose is the sugar found in malt balls, ice cream malts, and some malt flavored beers.

Sucrose, table sugar, is a disaccharide made of a glucose unit and a fructose unit linked by an α -1,2-glycosidic linkage.

As the number of monosaccharide units is increased it is possible to make various sugars and starches. These are numerous di- and trisaccharides found in nature. If the carbohydrates contains from four to ten monosaccharide units, it is designated as an oligosaccharide. If there are larger numbers of monosaccharides linked together, then the molecule is designated as a polysaccharide. Starches are polysaccharides containing as many as 30,000 glucose units connected by α-1,4-linkages with an α-1,6-linkage every 25 units or so.

Cellulose is a polysaccharide made of glucose units connected by β-1,4-linkages. This structural change makes it impossible for humans to digest cellulose.

28-9   Polypeptides and Proteins

Amino acids have both a basic group, an amine group, and an acidic group, the carboxyl group. There are 20 common amino acids that are found in proteins. Table 28-6 shows the structure of these 20 amino acids. Study it carefully. The carboxyl group on one amino acid can react with the amine group of a second amino acid to eliminate a water molecule and form an amide or **peptide bond** (Section 27-14). Using this reaction repetitively, it is possible to form dipeptides (two amino acids), tripeptides (three amino acids), and so forth up to polypeptides (many amino acids).

Example 28-2:   Draw the structure of the tripeptide, alanylalanylglycine.

Some di- and tripeptides have commercial and biological importance. For example, aspartame, the artificial sweetener Nutrasweet, is the methyl ester of a dipeptide.

The macrostructure of polypeptides, proteins, and their biological functions are determined by their structure. For polypeptides there are four levels of structure called the **primary, secondary, tertiary**, and **quaternary structures**. The order in which amino acids are linked together to make polypeptides is the **primary structure** of a protein. For instance, the tripeptides alanylalanylgycine, alanylglycylalanine, and glycylalanylalanine are all made of the same two amino acids yet each is a different tripeptide. They will each react differently and have slightly different properties. This is one of the key properties of polypeptides.

The **secondary structure** of proteins is the arrangement in space of the polypeptide backbone, ignoring the side chains that are attached to the amino acids. There are two basic shapes that are commonly found in proteins at the secondary structural level: α-helices and β-pleated sheets. See Figure 28-14 for the primary and secondary structures of myoglobin, an oxygen storing protein in muscles.

The **tertiary structure** of a protein is a description of the overall shape of the protein including side chains. For instance there are globular proteins which are "glob" shaped and generally water-soluble. Myoglobin is a globular protein. There are also fibrous proteins that are much more linear than globular proteins and water insoluble. The collagen in muscle tissue is a fibrous protein.

Quaternary structure of proteins is a description of the arrangement of subunits in the protein. For example, hemoglobin, the oxygen carrying protein in blood is made of four smaller proteins. Each of these smaller proteins must be arranged in a specific order to make the protein active.

If any of the four structures of a protein are changed or disturbed, the protein's function is changed or destroyed. A simple misplacement of an amino acid in hemoglobin's primary structure converts normal red blood cells into sickle cells. Disrupting the tertiary or quaternary structure of a protein can be accomplished with heat, solvents, heavy metals, or by agitation. This shape disruption is called **denaturing**. For example, surgical instruments are sterilized in an autoclave by heating them. This denatures the proteins of microorganisms and kills them.

28-10   Nucleic Acids

The biopolymers responsible for genetics are **nucleic acids**. The two nucleic acids are ribonucleic acid and deoxyribonucleic acid. Nucleic acids are made from two carbohydrates, five bases, and phosphate groups. The two carbohydrates found in nucleic acids are:

ribose   deoxyribose

The five bases in nucleic acids are

adenine   guanine

thymine   cytosine   uracil

A phosphate group has this structure

The arrangement of the carbohydrate, base and phosphate group in DNA to form a single nucleotide is:

The helix structure of DNA is formed by the phosphate group bonding to the -OH group of the deoxyribose in another nucleotide. This linkage is repeated over and over to form a single strand. H-bonding between the base pairs (Figure 28-18) on each strand attracts a second strand to the first strand forming the spiral stair case or double helix structure of DNA (Figure 28-19).

DNA is one of the largest molecules in nature. Human DNA has up to three billion base pairs and a molar mass of tens of billions. It is truly amazing that such a large molecule can be efficiently packed into the nucleus of each cell in our bodies.

Synthesis Question

Draw and name all of the structural isomers of the four carbon diols.

Group Question

Proteins are polyamides made from amino acids. Shown below are three simple amino acids. Draw all of the 3 amino acid long trimers that can be made from these three amino acids.

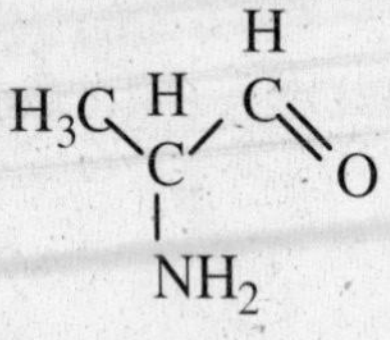

$$H_3C \underset{\underset{NH_2}{|}}{\overset{H}{\underset{}{C}}} \overset{H}{\underset{}{C}} \overset{}{\underset{}{C}} \!\!=\!\! O$$

alanine

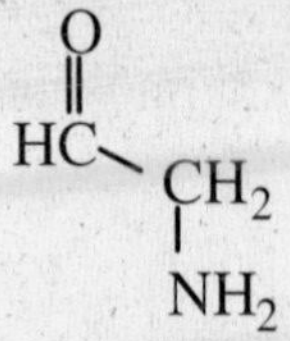

$$HC \overset{O}{\underset{\underset{NH_2}{|}}{\overset{}{\underset{}{CH_2}}}}$$

glycine

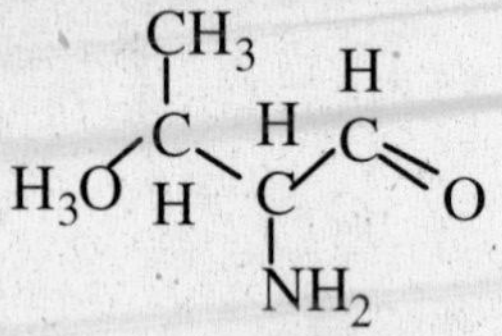

$$H_3C \overset{CH_3}{\underset{H}{\overset{|}{C}}} \overset{H}{\underset{\underset{NH_2}{|}}{C}} \overset{H}{C} \!\!=\!\! O$$

valine